Copyright © 2022 Independently published ISBN: 9798425492739

All rights reserved. This book or any portion thereof may not be reproduced without the written permission of the publisher, except for brief quotes in a review. Contact: Bernardo.Adeva@usc.es.

The Legrand Orange Book of Mathias Legrand is credited:
https://www.latextemplates.com/template/the-legrand-orange-book.
Cover design by Fausto Isorna: https://editorialgalaxia.gal/autor/fausto-isorna/.
Photography of Contents (Oberlausitzische Library of Görlitz, Germany) courtesy of:
https://imgur.com/gallery/BcsTm.
Photography of Chapter 1 (Uvac river, Serbia) licensed by:
Alberto Layo/shutterstock.com.
Photography of Chapter 15 courtesy of:
http://creativity103.com/collections/Vortex/slides/helterskelter-light.jpg
Photography of Chapter 16 courtesy of:
http://creativity103.com/collections/Vortex/thumbs/glowing-loops.jpg

Introduction

The course presented here is a brief, yet mathematically precise, introduction to Quantum Mechanics. It is addressed to students, general readers, or teachers, with some background in Calculus, classical mechanics and waves, and aware of complex numbers. Particularly to undergraduates following a course on Quantum Physics, in any Science or Engineering program.

A principal motivation of the book is seeking brevity and concision, trying to provide, in less than 150 effective pages and 39 figures, a full immersion in Quantum Mechanics, covering all key ideas and methodology. On this purpose, we try to provide a powerful framework that enables the reader to tackle problems of great diversity.

The introduction to Quantum Mechanics is based on the semiclassical framework, where Newton's classical mechanics and relativity are the reference points. Not only the radical conceptual differences that the latter show with respect to Quantum Mechanics are underlined, but also the smooth transition that is observed between them.

We have tried to avoid a double exposition of Quantum Mechanics by which the Planck constant phenomenology would be decoupled from the rest, purportedly embodied in a more rigorous conception.

Feynman's propagation is used as an axiomatic basis for Quantum Mechanics, for its unrivaled conceptual strength, completed with generally admitted ideas about the measurement problem.

From the teaching activity perspective, up to nine equivalent formulations of the *principles* of Quantum Mechanics have been discussed [1]. We give here special prominence, in addition to the aforementioned Feynman formulation, to that of Schrödinger's equation and wave function, that has been supplemented with a phase-space quantization approach. Despite being perhaps the most profound, we are not dealing with the matrix density formulation, of Von Neumann and Landau, which we find reasonably appropriate for a second course, somewhat more developed.

This course has been conducted over the last years at the university of Santiago de Compostela, as part of the graduate program in Physics, comprising an introduction to angular momentum and electron spin, that leads to the idea of fermion and boson systems. The analytical solution of a number of normalized cases of the Schrödinger equation is assessed, such as quantum tunneling and 3D oscillator. The present edition of the book includes improvements arisen from the 2022/23 course.

I would like to express my deep appreciation to Dolores Cortina Gil, Xabier Cid Vidal, Diego González Díaz, Beatriz Fernández Domínguez, Manuel Feijoo Rodríguez, Daniel Fernández Fernández, Pablo Baladrón Rodríguez, Gabriel García Jiménez, and Ramón Ruiz Fernández, for sharing with me the rewarding task of teaching and the development of the collection of problems related to the course, in interaction with the Santiago de Compostela students.

I am also grateful to Juan José Gómez Cadenas, for his praising comments to the first edition of the book. My deep recognition also goes to Marcos Seco Miguélez, for his valuable technical contributions, to Luciano Sánchez Aramburu's sensible corrections, and to Estrella Contreras Lamas, for her support. And not least to all Santiago students that, as David Castaño Bandín and Miguel Recuna Aranda, have with tireless attention contributed to free the text from inaccuracies and errata.

Santiago de Compostela, February 3, 2023.

---

[1] we recommend the excellent review article "Nine formulations of quantum mechanics" by D. F. Styler et al., https://aapt.scitation.org/doi/10.1119/1.1445404

Contents

1 THE PRINCIPLE OF LEAST ACTION 7

2 THE PLANCK CONSTANT 11
2.1 The observation through very short time intervals 13
2.2 The periodic motion 15
2.3 Atom sizes and the Bohr radius 16
2.4 The harmonic oscillator in 1D 20
2.5 Density of energy levels 21
2.6 The wave motion 23

3 THE FEYNMAN PROPAGATION 29
3.1 Exact propagation over a finite time 35

4 THE INSTANTANEOUS VELOCITY 37

5 THE SCHRÖDINGER EQUATION 41

6 THE WAVE FUNCTION 45

7 DE BROGLIE WAVE AND FOURIER TRANSFORMATION ... 49

8	MEAN VALUES AND UNCERTAINTY	53
9	THE UNCERTAINTY PRINCIPLE	57
10	EXTENSION TO THREE DIMENSIONS	61
11	EIGENSTATES AND MEASURABLE VALUES	63
12	THE STATIONARY STATES	67
13	THE BOHR FORMULA	75
14	QUANTUM MECHANICS IN A RELATIVISTIC FRAMEWORK	79
15	THE ANGULAR MOMENTUM	81
16	THE ANGULAR MOMENTUM EIGENSTATES	89
17	THE RADIAL SCHRÖDINGER EQUATION	101
18	THE BOHR MAGNETON	109
19	THE LINEAR POLARIZATION STATES	113
20	THE HYDROGEN ATOM	119
21	THE SPIN OF THE ELECTRON	127
22	THE PAULI EXCLUSION PRINCIPLE	135
23	THE PARTICLE IN A BOX	143
24	THE PIECE-WISE POTENTIALS	149
25	THE POTENTIAL BARRIER	155
26	QUANTUM TUNNELING	161
27	THE 3D HARMONIC OSCILLATOR	175

1. THE PRINCIPLE OF LEAST ACTION

Let us consider a moving body in one dimension, subject to a given potential energy $U(x,t)$. For instance an apple of mass m falling down to the ground from a tree branch, with uniformly accelerated motion. With $x(t)$ being its height from the ground, the kinetic energy equals $T = \frac{1}{2}m\dot{x}^2$, with potential energy $U(x) = mgx$. The Lagrangian is then defined by $L = T - U$. Starting from rest at an initial height x_1, the motion describes a trajectory in the (x,t) plane, given by the parabola $x = x_1 - \frac{1}{2}gt^2$.

Under the initial conditions given, the above trajectory is the *only* one that fulfils Newton's second law

$$-\frac{\partial U}{\partial x} = m\ddot{x}, \qquad (1.1)$$

or equivalently, the only solution to Lagrange's equation $\frac{d}{dt}(\frac{\partial L}{\partial \dot{x}}) - \frac{\partial L}{\partial x} = 0$. For any trajectory $(x(t),t)$, the action integral is defined as

$$S = \int_{t_1}^{t_2} L(x(t),\dot{x}(t),t)dt = \int_{t_1}^{t_2} (\frac{1}{2}m\dot{x}^2(t) - mgx(t))dt, \qquad (1.2)$$

where (t_1,t_2) represents any desired time interval where the action is to be calculated (e.g. from the instant the apple leaves the branch until it hits the ground). The action integral has the dimension of energy × time, for $L = T - U$ is an energy difference, that is multiplied by a time interval. In the International System of Units (SI) it is measured in Joule × second, $J \cdot s$.

It is known from as far back as the 18th Century that Newton's second law can be derived from a variational principle, the `principle of least action` stating that the action integral between the initial (x_1,t_1) and final (x_2,t_2) space-time points is an extremum [1] over the real trajectory $x(t)$ [2].

As a matter of fact, every differentiable function $x(t)$ differing from the trajectory $x(t) = x_1 - \frac{1}{2}gt^2$ will render a value of the action integral (1.2) larger than

$$S = \Delta t \left(-mgx_1 + \frac{1}{3}mg^2(\Delta t)^2\right).$$

We suggest to verify that indeed this expression is the action integral over the above parabola, after taking $t_1 = 0$ and $t_2 = \Delta t$.

Newton's second law can be derived from the least action principle in 3D by performing an infinitesimal variation of the action of a generic finite trajectory, and requiring it to be zero. This procedure is well known in Mechanics [3] and it leads to the Euler-Lagrange equations as a primary consequence.

Let us now see that a similar derivation can be attained in 1D, in a more direct way, by analysing the motion at a given point in the trajectory, through an infinitesimal time interval $\Delta t = t_2 - t_1$. Just assume that the body of mass m is moving from (x_1,t_1) to (x_2,t_2) in a very short time Δt, subject to the potential $U(x,t)$, and consider the position x occupied at the central time $t = (t_1+t_2)/2$ as represented in Figure 1.1.

The only intermediate point x allowed by the least action principle is precisely the one that meets Newton's second law $-\frac{\partial U}{\partial x} = m\ddot{x}$. Indeed, we can separately calculate the average velocities v_1 and v_2 in the first half (t_1,t) and in the second half (t,t_2) of the time interval. The acceleration at time t is then given by

$$\ddot{x} = \frac{v_2 - v_1}{\Delta t/2},$$

so Newton's second law can be expressed as

$$\frac{m}{(\Delta t/2)}\left[\frac{x_2 - x}{(\Delta t/2)} - \frac{x - x_1}{(\Delta t/2)}\right] + \frac{\partial U}{\partial x} = 0. \qquad (1.3)$$

---

[1] it can be shown that, for any potential, the action integral is a *minimum* for sufficiently short time intervals. In the most general case, it is either a minimum or a saddle point. The action can never be a maximum over the real trajectory.

[2] we take here Hamilton's formulation of the least action principle.

[3] see, for instance "Mechanics" Landau-Lifshitz, Vol.1, pag.2.

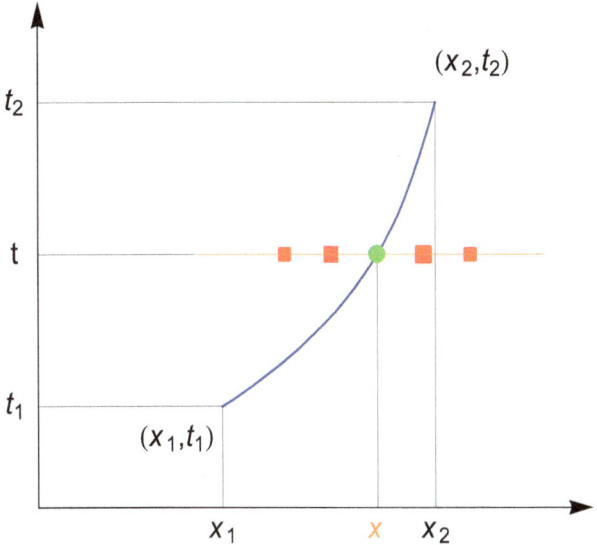

Figure 1.1: *Motion in the plane (x,t) according to different trajectories: that of minimal action (green) and others beyond Classical Mechanics (red).*

We can now calculate the action integral S over the interval Δt as a sum of two terms:

$$S = \int_{t_1}^{t_2} L\,dt = \int_{t_1}^{t_2}(T-U)dt = \bar{L}_1 \frac{\Delta t}{2} + \bar{L}_2 \frac{\Delta t}{2}$$

$$= \frac{\Delta t}{2}\left[m\frac{(x-x_1)^2}{2(\Delta t/2)^2} - U\left(\frac{x_1+x}{2}\right)\right] + \frac{\Delta t}{2}\left[m\frac{(x_2-x)^2}{2(\Delta t/2)^2} - U\left(\frac{x+x_2}{2}\right)\right] \quad (1.4)$$

where $U((x_1+x)/2)$ represents the average potential over the first half interval (similarly $U((x+x_2)/2)$). Obviously, the action will take different values for each intermediate point x that the particle may occupy at time $t = (t_1+t_2)/2$.

Yet the least action principle states that the only reachable point x is the one making the value of S *extremal* (minimum in this case), thus satisfying $\frac{\partial S}{\partial x} = 0$. It is easy to show, by differentiating expression (1.4) with respect to x, that formula (1.3) is readily obtained, that is equivalent to (1.1).

It should be made clear that, during the variational process considered above, the trajectory end-points (x_1,t_1) and (x_2,t_2) are to be held fixed.

In a time independent field ($\frac{\partial U}{\partial t} = \frac{\partial L}{\partial t} = 0$), the total energy $E = T + U$ becomes a *constant of the motion*. It is remarkable that in 3D a given point $(\mathbf{r}_2, t_2)$ may be reached from the initial point $(\mathbf{r}_1, t_1)$ through more than one real trajectory (all with extremal S).

Let us calculate in 1D the action integral over the points of a differentiable trajectory $(x(t), t)$ that, starting from the point (x_1, t_1), keeps a constant energy E [4]

$$S = \int_{t_1}^{t_2} dS = \int_{t_1}^{t_2} L(t) dt = \int \left(\frac{\partial S}{\partial t} dt + \frac{\partial S}{\partial x} dx \right) = -E \cdot (t_2 - t_1) + \int_{x_1}^{x_2} p(x) dx \quad (1.5)$$

The quantity

$$S_0 \equiv \int_{x_1}^{x_2} p(x) dx = S + E \cdot (t_2 - t_1) \quad (1.6)$$

is known in the literature as `reduced action integral`, also called characteristic action integral. It has the property of being *invariant* under any redefinition of the zero of the potential energy $U(x) \to U(x) + C$, due to the presence of an opposite sign term in $L = T - U$, thus being a *measurable* quantity in the laboratory. Because $p(x)$ and dx always have equal sign, it is a *positive-definite* magnitude. It plays a major role in the motion of particles and waves, and it has in Nature an *intrinsically oscillatory* character which we shall study next.

Be noted that S_0 takes different values depending upon the velocity of the observer, contrary to S, that is a *relativistic invariant*.

---

[4] it can be shown that $S(x_1, t_1; x, t)$ builds a function in the plane (x, t) which is called Hamilton's principal function, that fulfils the equations $\frac{\partial S}{\partial t} = -E$ and $\frac{\partial S}{\partial x} = p$, where p is the momentum at the point (x, t).

2. THE PLANCK CONSTANT

The reduced action S_0 being a measurable quantity, in the sense indicated above, it is conceivable to perform measurements of it, in units $J \cdot s$. This would require to determine the position and the kinetic energy of the moving body at multiple successive time slices, seeking to cause minimal disturbance to the trajectory. According to the laws of conventional mechanics, nothing prevents us to proceed in that way, and achieve a measurement of S_0 as accurate as we want, only limited by the precision of our experimental apparatus.

However the physical reality provides us a surprise. It is known from over a century ago that the reduced action is a *discrete* quantity, that appears to be *non zero* and take values which are integer multiples of a universal constant, the Planck constant, known today with 9 digits of precision. Its numerical value is close to $h = 6.626 \times 10^{-34} Js$, and it appears to be exactly the same, no matter the kind of energy that is represented by the potential $U(x,t)$, should it be electromagnetic, nuclear, electroweak or gravitational. Furthermore, it equally affects the motion of bodies of given mass m, and waves.

The true continuity of space and time at very short scales is an open question in physics, to which no established answer is available today. It is unknown how to address their possible discontinuity. However a discontinuity is well established in the small-scale observational processes, that refers unequivocally to the reduced action. Its impact in all branches of physics, and in the theory of knowledge in general, is huge, and we are going to perform a first discussion of the most important cases in what follows.

Taking the discontinuity of the reduced action as an additional postulate in physics, without any new ideas, would not be sufficient to create a new Mechanics of consistent and predictive nature. It would simply reflect reality. As we shall see next, such hypothesis is incompatible with the existence of differentiable trajectories, whereupon the laws of conventional mechanics lose their meaning. The problem of creating a new consistent and predictive Mechanics will be solved in Chapter 3. A *phenomenological statement* of the above fact, formulated under the limited terms of Classical Mechanics, goes as follows:

Principle of quantization of the reduced action
For every physical observation of a moving body, or an ensemble of them, subject to a force field, or of a wave, subject to the least action principle, both realized during a time interval Δt, the reduced action S_0 extended over Δt, appears to take values that are essentially integer multiples of the Planck constant. The quantization occurs in the form $S_0 = (n+\alpha)h$, where $n = 1, 2, \ldots \infty$, and $\alpha > -1$ is a constant, specific to each problem.

The above principle is universally applied to systems with an arbitrary number of degrees of freedom N, whether they be relativistic or non relativistic. It holds for integrable systems, where $N-1$ constants of motion exist other than the total energy, as well as for fully chaotic systems where only energy is conserved. A modern view reveals it is an excellent approximation, even if not completely exact. As we shall see in Feynman's theory, its lack of complete accuracy is a consequence of the fact that not only the classical trajectory contributes to the motion, but *all* possible trajectories.

As an approximate expression, the principle as such will be derived easily from Feynman's formulation of Quantum Mechanics, in Chapter 12. The constant α is related to focusing properties of the classical trajectories in periodic motion, and it can actually be interpreted precisely in Classical Mechanics. We shall not elaborate on this here [1], but rather focus on the development of the new mechanics.

---

[1] α is meaningful for integrable systems in periodic motion, and it is additive for each irreducible closed circuit around the *invariant tori in phase-space* that characterize this kind of systems. For each degree of freedom i, and assuming n runs from zero $n = 0, 1, \ldots \infty$, it takes the value $\alpha_i = \mathcal{M}_i/4$ with $\mathcal{M}_i$ being a positive integer, called the Maslov index.

Some comments are still in order, about the action quantization principle:

- given the fact that differentiable trajectories do not actually represent the real motion, it is difficult to verify its validity in a *direct* way. If we try to perform a measurement of S_0, it is found that the *precision* required to test the hypothesis (better than h) can never be attained in practice, as we shall see in Chapter 9, because the system will *necessarily* get perturbed.
- one way to test the principle, regarding periodic motion, is to measure the *energies* belonging to the sequence $n = 1, 2, \ldots \infty$, giving up any attempt to observe the trajectory at the same time. It turns out that these can indeed be measured with unlimited precision. Independently, for any kind of motion, it is always possible to measure the energy and the time elapsed during the measurement, by processing signals from the moving body with clocks of given precision.
- the trajectories where the S_0 quantization rule is accomplished, in periodic motion, are *assumed* to be differentiable, and they should be understood as the *closest approach* to the real motion that can be formulated under the terms of Classical Mechanics.

Given the small value in $J \cdot s$ that the Planck constant takes in Nature [2], the domain of physics splits up into two sectors, with a continuous transition between them: either its non zero value is imperceptible, or its effects produce a noticeable difference. The former case happens when, once the reduced action S_0 is evaluated for a given problem, it appears to be $S_0 \gg h$. We shall call it the `classical limit`, where we would expect full validity of the conventional laws of mechanics. The latter occurs quite the opposite, when $S_0 \simeq h$, and it is known as the `quantum limit`, which we shall deal with in the following.

2.1 The observation through very short time intervals

Let us observe a moving body of kinetic energy $T = mv^2/2$ for a time interval $2\Delta t$ (Δt being the approximate precision of its time location). The discrete character of the reduced action $S_0 = nh$ ($n = 1, 2 \cdots \infty$) forces that the mere observation will induce an increase of its kinetic energy $\Delta T \geq 0$, provided that Δt is short enough.

---

[2] both the Joule and the second are units bound to our *human* evolutionary observational scale.

Indeed: $S_0 = 2(T+\Delta T)2\Delta t = nh \geq h$. If $\Delta t \ll h/(4T)$ then the first term can be neglected, thus [3]:

$$\Delta T \geq \frac{h}{4}\frac{1}{\Delta t} \gg T,$$

and the kinetic energy becomes infinitely large for $\Delta t \to 0$. The above phenomenon of obtaining $\Delta T \neq 0$ is called quantum fluctuation [4].

Hence we see that the critical time for quantum fluctuations to become significant (for the free motion $E = T$) is:

$$\Delta t \lesssim \frac{h}{E} \qquad (2.1)$$

We suggest checking that the critical observation time for a tennis ball of 100g moving at 50 km/h would be $\Delta t \sim 6.9 \times 10^{-35} s$, which is less by many orders of magnitude than the exposure time of any photographic camera ($\Delta t \geq 1ms$), or than the light collection time for any electronic device based on photo-sensitive cells ($\Delta t \geq 1ns$). Nonetheless if we were able to "take a photo" of the moving body during a time interval of $10^{-35} s$, we would undoubtedly observe fluctuations, and we could see that the motion actually takes place in a zigzag manner, on that scale.

That is to say, the time location Δt of a moving body in the laboratory entails an approximate energy increase by $h/\Delta t$. This energy can be regarded as a necessary expenditure to achieve that location, or as a simple manifestation of it. In either case we cannot maintain the existence of an instantaneous velocity, as determined through the *observational process* of taking the limit

$$v = \lim_{\Delta t \to 0} \frac{\Delta x}{\Delta t},$$

and the above limit does not exist (it is infinite). Therefore moving bodies do not follow differentiable trajectories. It just appears to us that they do so, due to the fact that our senses or measurement devices are not able to verify sufficiently short time intervals.

However, the periodic motion of electrons inside atoms, molecules, or crystal lattices, as well as that of protons and neutrons inside atomic nuclei, entails a short time location, defined by their revolution period, thus being entirely governed by quantum fluctuations.

---

[3] to estimate $\Delta T/T$ precisely it becomes necessary to know the details of the *spatial* location of the body (e.g. fluctuations of its center of gravity), and we cannot in general assume that moving bodies are *mathematical* space points (as generally assumed in Mechanics).

[4] only for $\Delta t \ll h/(4T)$ it is garanteed that ΔT is positive. For $\Delta t \lesssim h/T$, softer quantum fluctuations take place with ΔT being either positive or negative, subject to the discrete values $n = 1, 2, 3,$ or 4.

Relativistic consideration

In giving up the existence of differentiable trajectories, we have restricted ourselves to the consideration of the non relativistic kinetic energy. Yet it is clear that for sufficiently short observation times, the increase of the kinetic energy necessarily takes every moving body to the relativistic limit, and its velocity will approach the light velocity c. An instructive exercise for the interested student is to show the above by taking the relativistic action S for the free motion [5], which implies the reduced action $S_0 = S + mc^2\gamma\Delta t$, where γ is the Lorentz factor. It is then straightforward to find out that, assuming $S_0 \sim h$, the time location Δt necessary for the particle to reach a velocity very close to c is given by $\Delta t \sim h/(mc^2)$, where m is the particle's rest mass. Note that such time location is in general *much shorter* than the one necessary to produce significant quantum fluctuations.

If we were to achieve the above time location with particles like atoms, molecules, or atomic nuclei, it would mean such an extremely short time, that those systems would lose their integrity, and its consideration would be of no relevance to their study. For an electron, of much lighter mass m_e, the time location $\Delta t \sim h/(m_e c^2)$ is still too short as to be relevant for the study of ordinary matter, despite having the interesting implication of producing electron-positron pairs, wherever that location would be realized.

2.2 The periodic motion

Given the fact that a great deal of motions observed in nature are of periodic type (particularly in the microscopic domain) let us see what the condition is for such a motion to be in the *quantum* limit.

Let us assume for instance a closed orbit in three dimensions, with a central potential of the type $U(r) = \beta r^d$ ($d > -2$). According to the virial theorem, the mean values of the kinetic ($\bar{T}$) and potential ($\bar{U}$) energy, extended over one period [6] Δt, are exactly related by the expression $\bar{T} = \frac{d}{2}\bar{U}$.

This theorem allows to perform the exact calculation of the action integral over one cycle

$$S = \int_0^{\Delta t} L\,dt = \bar{L}\,\Delta t = (\bar{T} - \bar{U})\,\Delta t = \left(\frac{d-2}{d+2}\right) E\,\Delta t\ . \qquad (2.2)$$

---

[5] the relativistic action for the free motion is $S = -\int_0^{\Delta t} mc^2\sqrt{1-v^2/c^2}\,dt$.
[6] for central potentials of the type indicated above, only the cases $d = -1$ and $d = 2$ give rise to trajectories which are always periodic. However, in all other cases it would make perfect sense to replace the period with the time interval Δt over which r completes a cycle between (r_{min}, r_{max}), called here partial period.

The energy $E = \bar{E} = \bar{T} + \bar{U}$ remains a constant of the motion in the above integral, and the reduced action S_0 can be readily obtained from S by adding the term: $S_0 = S + E\Delta t = (2d/(d+2))E\Delta t$. The quantum limit is thus reached when $S_0 \sim h$. In this kind of periodic motion, in one or more dimensions, there exists a 1–1 relationship between the period Δt (or the partial period) and the energy E of the orbits, the only exception being the harmonic oscillator ($d = 2$), where the period turns out to be independent of the energy.

Therefore, once the energy E and the period Δt are given, we know precisely whether the motion takes place in the quantum limit or not, by simply evaluating their product. In the quantum limit it will not be even possible to talk about trajectories, and the motion will happen erratically, with quantum fluctuations acting strongly over each cycle.

2.3 Atom sizes and the Bohr radius

In order to understand the finite size of atoms, let us begin with the simpler case of Hydrogen. Consider the problem of an electron moving under the electric attraction force of a proton (Z protons in general). The interaction potential is given by Coulomb's law $U(r) = -\beta/r$, with $\beta = \frac{Ze^2}{4\pi\varepsilon_0}$, where e is the electron charge and ε_0 the vacuum permittivity [7].

According to Newton's laws, the motion takes place in a plane, and for energies $E < 0$ the solutions are the well-known elliptic orbits in which one of the particles moves around the center of mass, located at one of the foci. Furthermore, Kepler's third law

$$\Delta t = \frac{\pi}{\sqrt{2}} \frac{\beta m^{1/2}}{|E|^{3/2}} \equiv \frac{2\pi}{\omega} \tag{2.3}$$

establishes a precise relationship between the orbit energy and its period Δt, related to the angular frequency through the expression $\Delta t \equiv 2\pi/\omega$.

Note that m represents here the reduced mass of the two-body system (electron and nucleus) $m = m_1 m_2/(m_1 + m_2)$, which is quite close to the electron mass m_e. As we have seen, according to the virial theorem ($d = -1$ in this case), the action takes the value $S = -3E\Delta t$ and the reduced action $S_0 = -2E\Delta t$.

---

[7] this potential also describes the gravitational attraction between two masses m_1 and m_2, with $\beta = G_N m_1 m_2$, and the ensuing physics analysis entirely holds in this case as well.

2.3 Atom sizes and the Bohr radius

The quantum limit will occur when the action S_0 falls to the level of h. From Kepler's third law, on substitution of expression (2.3) in the above equation, it happens when the orbits acquire very small radii, high frequencies, and negative energies with high absolute value.

The reduced action quantization principle is then applied: $S_0 = nh$ with $n = 1, 2, \ldots \infty$, with a constant α that is effectively zero in this case [8].

Hence squaring (2.3), the above principle sets a filter on the allowed electron energies, which become quantized in the form

$$-E_n = \frac{1}{2}\frac{m\beta^2}{\hbar^2}\frac{1}{n^2} = \frac{1}{2}\frac{mZ^2 e^4}{(4\pi\varepsilon_0)^2 \hbar^2}\frac{1}{n^2}. \qquad (2.4)$$

Their value for $n = 1$ and $Z = 1$ ($\hbar \equiv h/2\pi$) corresponds to the minimum energy of an electron in the electric field of a proton, and the progression of the energies towards zero for $n \to \infty$ is depicted in Figure 2.1.

The electron binding energy in a Hydrogen atom is known with great precision (better than one part in 10^8), from the wavelengths of its emission lines (Balmer series), and it is called in the literature Rydberg energy E_y. The following numerical value is obtained from (2.4), when using precision values of the fundamental constants involved

$$-E_1 = \frac{1}{2}\frac{m\beta^2}{\hbar^2} = 2.180 \times 10^{-18} J, \qquad (2.5)$$

that is in excellent agreement with the measured value of E_y, from which it differs by a relative amount of 0.42×10^{-3}. This is attributed to having ignored the electron magnetic moment, to the lack of precision of the non relativistic calculation, to the finite proton mass, to the proton magnetic moment, and to the vacuum polarization.

Such state of minimum energy, which arises from quantum fluctuations in binding potentials, is generically called ground state . For Hydrogen its magnitude is not infinite, E_1 being exactly equivalent to the energy acquired by an electron through a potential bias of 13.6 Volts (13.6 eV).

---

[8] to be more precise, the quantization condition in this case is: $S_0 = (n + \mathcal{M}/4)h$, with the Maslov index $\mathcal{M} = 4$. In 2D we have $\mathcal{M} = 2$, meaning two conjugate points on the ellipse sitting on a straight line through the secondary focus. But this ellipse must be embedded in 3D, and two more conjugate points appear, sitting on the straight line through the main focus (one of them is repeated). See M. Gutzwiller, "Chaos in Classical and Quantum Mechanics", Springer 1990. The general rule, with several degrees of freedom, is that at least one of them has $\mathcal{M} \neq 0$ and n actually runs from zero $n = 0, 1, \ldots \infty$.

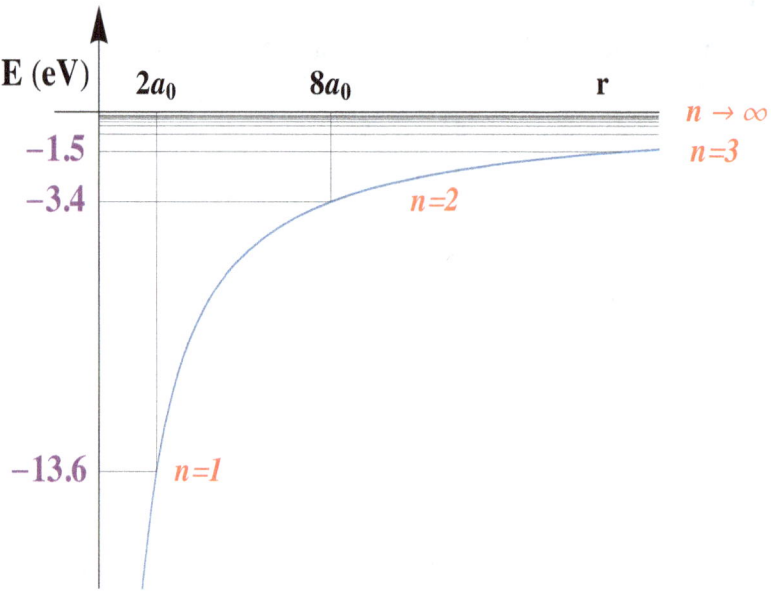

Figure 2.1: *Coulomb potential in the Hydrogen atom showing the discrete energy states that arise when the reduced classical action S_0 becomes an integer multiple of h. The lowest level E_1 defines the Bohr radius $a_0 = \beta/(-2E_1)$. For $n \to \infty$ the energies approach continuous values.*

As we have shown, E_1 is proportional to the coupling constant squared β^2 (the "intensity" of the interaction), to the mass m of the *fluctuating* particle, the electron, and inversely proportional to the square of the Planck constant. Should h be zero, the electron energies would become infinitely negative ($E \to -\infty$), and in fact no other principle of mechanics would prevent this from happening, no matter the initial conditions [9].

If we think in terms of the classical trajectory, the electron returns to the same position after completing one cycle, which implies a time location for the extent of the period ($\sim 10^{-16} s$).

Since the above time is close to the ratio h/T (T being the electron kinetic energy), quantum fluctuations become important. They are actually responsible for the kinetic energy itself, causing the motion around the proton to be erratic. At the ground state, the motion does not resemble any kind of elliptical trajectory, neither does it take place on a plane.

---

[9] according to Maxwell's equations of Electromagnetism, every charge e subject to an acceleration a radiates energy with power $\mathcal{P}$ in Watts given by **Larmor's radiation law**: $\mathcal{P} = \frac{e^2 a^2}{6\pi\varepsilon_0 c^3}$ in the nonrelativistic limit. The electron energy loss would cause a very rapid fall onto the proton ($\sim 0.1\, ns$ time scale). We suggest to assess and solve the related differential equation, assuming circular orbits.

However the *most probable distance* between the electron and the proton is well defined, and univocally determined by its energy E, according to the expression $a = \beta/(-2E)$ that gives the major axis in Kepler's motion.

The experimental data indicate that the 3 components of the orbital angular momentum L are zero at the ground state, so that the classical ellipse degenerates into a straight line, and the radial distance is oriented at random in 3D by quantum fluctuations. Actually the atom acquires a spherical shape, where the most probable radius is precisely defined, being known in the literature as the **Bohr radius** a_0. It is predicted to be

$$a_0 = \frac{\beta}{-2E_1} = \left(\frac{Ze^2}{4\pi\varepsilon_0}\right)\frac{1}{-2E_1} = \frac{\hbar^2}{Z}\frac{4\pi\varepsilon_0}{me^2},$$

which after numerical evaluation with precision values of m_e, ε_0, and e, for $Z = 1$, yields

$$a_0 = 0.529 \times 10^{-10} m = 0.529 \text{Å} = 0.0529 \ nm = 52.9 \ pm,$$

where the angstrom (Å) is an ad hoc unit $1\text{Å} \equiv 10^{-10} m$ sometimes used in atomic physics. Although too small to be observed with an optical microscope, due to light diffraction, it can nonetheless be observed with more powerful experimental techniques.

The existence of discrete energy levels in Hydrogen according to (2.4), in correspondance with the squares of the natural numbers $n = 1, 2, \ldots \infty$, is well established experimentally through the wavelengths of the light emitted by the electron, when the atoms are subject to thermal collisions with an equivalent temperature of tens of thousands of degrees.

Note how in the limit $n \to \infty$ the allowed energies seem to recover continuous values, for the differences $|E_{n+1} - E_n|$ become very small with respect to $|E_n|$, as it can be appreciated in Figure 2.1. This is to be expected in the classical limit.

The state of minimum energy that arises in the atom can be understood as a balance between two opposite forces: the energy loss due to Larmor's radiation law, that tends to bring the electron closer to the proton, and the kinetic energy increase due to quantum fluctuations, which become very large at small distances (very short periods), that tends to separate it from the proton. The former arises from the electron acceleration in the Coulomb field. Equilibrium is reached at a most probable distance that is no other than the Bohr radius.

One would think of an analogy with Kepler's classical motion, where such a balance also occurs between the centrifugal force and the attraction force. However, a common *misconception* is to be made straight: the ground state energy of the Hydrogen atom does not originate from the centrifugal barrier, as the orbital angular momentum of the electron is known to be *exactly* zero.

It is thus seen that the size of the Hydrogen atom (and similarly of all other atoms) is actually determined by the Plank constant: `if it were zero, all atoms would be infinitely small`. So the *atomic structure* of matter originates from quantum fluctuations.

2.4 The harmonic oscillator in 1D

Every potential well in one dimension $U(x)$ generates, for a particle of mass m, a periodic motion between its two turning points. The parabola $U(x) = \frac{1}{2}kx^2 = \frac{1}{2}m\omega^2 x^2$ generates the so-called harmonic oscillator, that shows the remarkable property that the period (frequency) of the oscillations $\Delta t \equiv 2\pi/\omega$ becomes independent of the total energy E. The latter is only determined by the amplitude A, $E = \frac{1}{2}kA^2$.

All branches of physics profusely refer to harmonic oscillators. We might think that the energy (amplitude) of an oscillator would be as small as we want, but nothing further from the truth. The quantization principle of the reduced classical action prevents that from happening.

Given the general solution for the trajectory $x(t) = A\cos(\omega t - \phi)$, we suggest to verify that a zero value is obtained for the action integral ($S = 0$) when extended over the period Δt. Nonetheless the reduced action is $S_0 = S + E\Delta t = 2\pi E/\omega$, and the action quantization principle implies in this case $S_0 = (n + \mathcal{M}/4)h$ with $\mathcal{M} = 2$ [10] and $n = 0, 1, \ldots \infty$.

The above renders the energy levels of the harmonic oscillator

$$E_n = \left(n + \frac{1}{2}\right)\hbar\omega \qquad n = 0, 1, 2, \ldots \infty, \tag{2.6}$$

which is one of the most far-reaching and best tested results in Physics. It means that no oscillator can gain or transfer energy by an amount smaller than $\hbar\omega$. This quantity is called in the literature quantum of energy, and grows *linearly* with frequency, as has been shown.

---

[10] in every potential well in 1D, the Maslov index $\mathcal{M}$ equals the number of turning points, namely two.

The quantum was originally conjectured by the German physicist Max Planck in 1900, as refering to the atomic oscillators that belong to the walls of every cavity in equilibrium with its radiation. The above expression for E_n is also telling us that no oscillator can vibrate with energy less than $1/2$ of the quantum. This is the so-called zero-point energy which completely revolutionized the low-temperature thermodynamics in the early 20th century.

For a 3D oscillator (as atoms in a crystal lattice are, for instance), the minimum energy is actually $(3/2)\hbar\omega$, where ω represents the frequency of sound in that body. One mole retains an energy $N_A(3/2)\hbar\omega$, even at the absolute zero of temperature ($T \to 0$), N_A being Avogadro's number.

2.5 Density of energy levels

We have seen, in two simple and most characteristic Hamiltonian systems, how the reduced action quantization generates discrete energy levels. Indeed, this is actually a general feature, that is independent of the nature of the system. As will be seen below, the discretization of the energy is related to the fact that the phase-space is itself quantized.

Consider a general Hamiltonian $H(q,p)$ with N degrees of freedom, having generalized coordinates q and momenta p in a $2N$-dimensional phase-space. The reduced action quantization over *periodic* trajectories of the system is expressed by the closed line integral

$$S_0 = \oint \boldsymbol{p}d\boldsymbol{q} = \sum_{i=1}^{N} \oint p_i dq_i = \sum_{i=1}^{N} S_0^i = \sum_{i=1}^{N} (n_i + \alpha_i)h ,$$

where $\oint p_i dq_i = \iint dq_i dp_i = (n_i + \alpha_i)h$, from Green's theorem [11].

Therefore the area enclosed by the trajectory in the (q_i, p_i) phase-space is approximately given by a multiple n_i of an elementary pixel $\delta q_i \delta p_i$ of area h, as illustrated in Figure 2.2.

Hence every volume in $(\boldsymbol{q}, \boldsymbol{p})$ space can be approximated by an integer multiple of the elementary $2N$-cube $(\delta q_1 \delta p_1)(\delta q_2 \delta p_2) \cdots (\delta q_N \delta p_N) = h^N$. Which means the phase-space becomes itself partitioned in elementary pixels of volume h^N.

---

[11] the line integral should be taken *clockwise*, since $p_i dq_i$ is always nonnegative.

In Classical Mechanics, the total volume of the phase-space occupied by all trajectories with energy below a given value E, is defined by the following multiple integral

$$\Omega(E) = \int d^N q \left(\int \theta\big(E - H(q,p)\big) d^N p \right)$$

where H is the Hamiltonian of the system and $\theta(x)$ is the unit step (or Heaviside) function [12].

Having the above partitioning in mind, we define the average *number of energy levels* as $\langle \mathcal{N}(E) \rangle \equiv \Omega(E)/h^N$, with $\langle d\mathcal{N}/dE \rangle \equiv (d\Omega(E)/dE)/h^N$ being the average *density* of levels. The average quantities $\langle \mathcal{N}(E) \rangle$ and $\langle d\mathcal{N}/dE \rangle$ illustrate the fact that, for every physical system, the reduced action quantization creates in general a *finite* number of levels in a given energy interval [13], that can be calculated just from the knowledge of Planck's constant, and of the classical phase-space volume $\Omega(E)$. This calculation is actually quite accurate, if we apply it to energy intervals $(E, E + \Delta E)$, with ΔE small in comparison with E, but large enough as compared with the mean energy spacing $\langle d\mathcal{N}/dE \rangle^{-1}$.

Thus we see how every physical system, that we have assumed to be confined in periodic motion, is subject to the presence of discrete energy levels, which is in sharp contrast with Classical Mechanics, where the energy is allowed to take continuous values. It is revealing that discretization holds in the case we previously defined as the *classical limit*, showing that Classical Mechanics is *never actually retrieved*, even in that limit.

Let us finally mention an interesting curiosity for the keen student, whose details would take us far from the scope of this course. For classically chaotic systems, where no constants of motion exist other than the total energy E, the quantized level spacings appear to be remarkably *regular*.

In fact, the number of levels $\mathcal{N}(E_q)$ below the energy E_q, as a function of $E_q \in (E, E + \Delta E)$, fits well to a straight line, and the *mean square deviation* of the fit turns out to be much smaller for chaotic systems than it is for integrable systems [14].

---

[12] we propose as an exercise to show that for a 1D oscillator $\Omega(E) = 2\pi E/\omega$, and for an ensemble of N independent oscillators with Hamiltonian $H = \sum_i^N p_i^2/2m + (1/2)m\omega^2 q_i^2$, $\Omega(E) = (2\pi E/\omega)^N/(N\Gamma(N))$, with $\Gamma(N) = (N-1)!$. To prove it, use the volume of the n-ellipsoid $V_n = (a_1 \cdot a_2 \cdots a_n)(2/n)\pi^{n/2}/\Gamma(n/2)$.

[13] for particular choices of the energy interval, the number of levels may become infinite, always being an enumerable set.

[14] see for further details on this point "Semiclassical theory of spectral rigidity", M.V. Berry, Proc. R. Soc. London A400, p. 242 (1985).

2.6 The wave motion

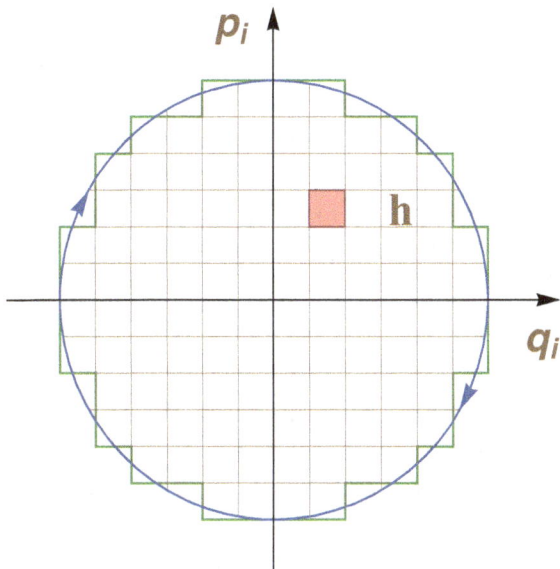

Figure 2.2: *Periodic trajectory of one degree of freedom i, having coordinate q_i and generalized momentum p_i. The phase-space volume is pixelated in multiples of h.*

In the latter case, where $N-1$ constants of motion exist, the level positioning surprisingly occurs at random in general, with respect to the mean energy spacing. Both phenomena are understood [15], and the former is known in the literature as *quantum level repulsion*. It has been extensively observed in classically chaotic systems with a large number of freedoms N, including compound-nucleus resonances, heavy atoms and molecules.

2.6 The wave motion

The realm of physics primarily consists of two kinds of objects: moving bodies of given mass m, and waves. Yet the discontinuous nature of the reduced action affects both of them equally.

Waves are objects that describe the periodic propagation of physical magnitudes in an a priori continuous manner across space and time. In particular, all *fields* in physics are actually waves, when their sources are subject to acceleration.

---

[15] a detailed discussion on this issue will be found in the book M. Gutzwiller, "Chaos in Classical and Quantum Mechanics", pp. 270-278, Springer 1990.

For the sake of simplicity, we restrict ourselves to one dimension, even if their natural domain are the three space dimensions. Their most general mathematical definition is a propagation amplitude in the form $Ae^{i(kx-\omega t)}$, where A is the physical magnitude that propagates, measured in appropriate SI units, ω is the time periodicity, or angular frequency (units s^{-1}) and k is the spatial periodicity or number of waves per unit length (m^{-1}). Both periodicities are always related by a function $\omega = \omega(k)$, which is called dispersion relation. There exists in physics a great diversity of waves, either propagating in a material medium or in a vacuum, where nearly all conceivable functions are realized as $\omega(k)$. This function must in general be determined experimentally. The utilization of complex numbers to describe waves is convenient, but not essential, since all their properties could equally well be assessed with the cosine function.

When $\omega(k)$ is the linear function $\omega = vk$ we talk about a non dispersive wave, with v being its propagation velocity. In general the group velocity $v = d\omega/dk$ defines the propagation of wave pulses.

Waves carry energy (and thereby information) across the space. At each space point they have an energy density $\mathcal{E}$ per unit volume. They also have a Lagrangian density $\mathcal{L}$ (per unit volume) [16], that governs, through the least action principle, the partial differential equations that the magnitude A may be subject to fulfil. When the waves propagate in a material medium made of atoms or molecules, these act as carriers just because they vibrate harmonically around their equilibrium positions. For that reason the wave inherits from the harmonic oscillator its essential properties.

Let us consider a wave with frequency ω, enclosed in a rectangular box of volume V, whose dimensions are integer multiples of half the wavelength.

The reduced action, extended over one cycle $t_2 - t_1 \equiv 2\pi/\omega$, takes the value

$$S_0 = \int_{t_1}^{t_2} \left(\int_V \mathcal{L}(x,\dot{x},t) d^3x \right) dt + E \cdot (t_2 - t_1),$$

where E is the integral of the energy density $\mathcal{E}$ over the volume V.

Both in the case of a continuous distribution of oscillators and of an electromagnetic wave, the first term turns out to be zero [17]. Keep in mind that we actually have three waves, on each of the coordinate axes X,Y,Z.

---

[16] for instance, the Lagrangian density for the electromagnetic field, in absence of currents and charges, is given by $\mathcal{L} = \frac{1}{2}(\varepsilon_0 E^2 - \frac{1}{\mu_0} B^2)$.

[17] as in the case of the harmonic oscillator, this is simple to work out, and in both cases it is related to the integral of a cosine function over an integer number of periods.

2.6 The wave motion

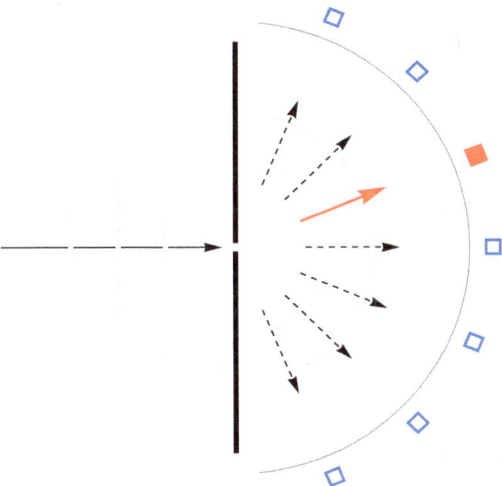

Figure 2.3: *When a low intensity wave is diffracted, only one of the detectors will register the quantum, the rest will not fire. We cannot predict which one of them will fire.*

The principle of the reduced action quantization, using the same results as for the harmonic oscillator, leads to the conclusion that the energy in the box is quantized in the same way it is for the oscillator

$$E_n = \left(n + \frac{3}{2}\right)\hbar\omega \qquad n = 0, 1, 2, \ldots \infty. \tag{2.7}$$

If we open a hole in one of the box's walls, the wave will propagate out with a certain velocity, and the energy transferred towards a given detector will be quantized in full units of the quantum $\hbar\omega$, which is directly proportional to the frequency. This happens irrespective of whether the wave propagates in a medium or in a vacuum, or of the physical nature of the propagating magnitude A. The existence of a well defined frequency is what matters. Thus the *transferable* energy of any wave amounts to an integer number of quanta $E = n\hbar\omega$, with $n = 1, 2, \ldots \infty$.

However, the wave enclosed in the box of volume V still has a non transferable energy $\frac{3}{2}\hbar\omega$ which is stored inside and not carried away with the wave. If the wave propagates in a vacuum, it is natural to think that all possible frequencies must contribute to its zero-point energy.

The fact that summing over all frequencies, for the wave propagating in a vacuum, provides an infinite energy (as $\omega_n \to \infty$) should not be surprising, given that infinitely short time intervals come into play [18]. Yet if the wave propagates through atoms or molecules, an upper limit to its possible frequency arises, when the minimum wavelength equals the atom sizes, and no infinities can possibly emerge.

Let us consider a wave that carries a certain power per unit area $\mathcal{P}$ through its wave front (in Wm^{-2}), impinging on a normal surface of area A_S. For example, an electromagnetic wave. We shall underline four essential aspects related to the detection of quanta in the laboratory, in the limit where the wave power is very weak, and the individual signals from quanta are detectable:

- The transfer of the quantum is instantaneous, sudden, and not gradual. The time at which the transfer takes place, counted from the instant the wavefront arrives to the surface, is unpredictable, within the time interval $\tau = \hbar\omega/(\mathcal{P}A_S)$ associated with the *average* time distance between the arrival of quanta [19].
- To all effects, the wave behaves as a collection of particles which travel with it, although they are *not individually synchronized* with it.
- The random character of the quantum extends to the spatial direction of the energy transfer. Take the example of a wave that is diffracted through a small hole, and set an ensemble of detectors uniformly over a spherical surface centered in it. It is *unpredictable* which detector will fire and which will not. This is illustrated in Figure 2.3. A relevant example can be found in the direction of emission of a photon from an atom, or from an atomic nucleus, within the 4π solid angle.
- The idea of instantaneous power (in Wm^{-2}) becomes meaningless in the limit of very low intensity waves. Obviously, we have an infinite instantaneous power for the absorption of a quantum, if we divide a finite energy over *zero* time. The wave power must then be understood as an *average* power. What happens here is *not dissimilar* to the failure we encountered in the definition of the instantaneous velocity of a particle of mass m.

---

[18] the zero-point energy has been indirectly observed in precision experiments (S. K. Lamoreaux, PRL 78, 5 (1997)), in the case of the electromagnetic field. The field fluctuations in a vacuum generate an attractive force between two very close parallel metallic plates, known as **Casimir force** in the literature.

[19] the probability to observe n quanta in a time interval Δt is actually given by a **Poisson distribution** $P(n) = e^{-\lambda}\lambda^n/n!$ with $\lambda = \Delta t/\tau$.

2.6 The wave motion

Just as the waves they represent, the quanta transfer not only energy, but also momentum, and in many cases, angular momentum. There is always a matter source where quanta can be absorbed into or emitted from. Cases of major importance in physics are:

- the photon for electromagnetic waves, the source being the electric charges.
- the phonon for sound waves in solids or liquids, the source being atoms or molecules in a spatial lattice.
- the gluon for color waves originating from the quarks. These waves (chromoelectric and chromomagnetic fields), are quite similar to the electromagnetic field, and they transmit the strong interaction between quarks that is behind the nuclear force. They are responsible for the existence of protons and neutrons.

The previously described properties concerning the detection of quanta have been observed in the three cases above, with remarkably high precision in the photon case. It is clear that the existence of quanta introduces a random element in the energy transmission by waves.

A surprising consequence of the presence of the Planck constant in physics is, as we have just seen, that both moving bodies of mass m and waves acquire similar properties, for they share a common random nature, which in neither case was to be expected. In the absence of measurements, both moving bodies of mass m and waves propagate simultaneously through many different paths, as we shall study next. Furthermore the realization of an experimental measurement forces the reduction of all paths to only one.

3. THE FEYNMAN PROPAGATION

As stated earlier, the discrete and *non zero* character shown by the reduced action $S_0 = nh$ ($n = 1, 2 \cdots \infty$), requires that any measurement attempted on a particle for a time $\Delta t \ll h/E$ will generate an increase of its kinetic energy that is not reconcilable with a differentiable trajectory.

If we try to imagine the motion as a succession of small time intervals, we should not be surprised that accepting the discrete character of the reduced action (as originally encountered by Planck) entails a great *conceptual transformation* in physics, which can be summarized as follows: the motion for $\Delta t \lesssim h/E$ takes place in a non deterministic way, such that the observed position of a particle at a given time, cannot be inferred with certainty from its position and velocity at an earlier time.

Such statement may seem astonishing, and indeed it is against what the theory of differential equations tells us, where the specification of the initial conditions (position and velocity) suffices to determine the unique solution for the motion at any later time t. However, the above theory is based upon the *differentiability* of the trajectories.

It is therefore understood why a suitable description of physics that includes the motion of atoms, molecules and elementary particles, compels us to abandon its deterministic character. On the other hand, it is clear that any formulation given along these lines must also recover the laws of classical motion, of deterministic nature, which we know describe with great precision the motion for large observation time intervals $\Delta t \gg \frac{h}{E}$.

Let us see how such formulation can be achieved in a *rigorous* manner. For this purpose, we focus on the one dimensional motion, and come back to the discussion of Chapter 1, for a particle moving during a short time interval $\Delta t \to 0$, subject to a potential $U(x,t)$. The real motion is thought to "chose", among the infinite intermediate positions x between x_1 and x_2, the one satisfying second Newton's law $-\frac{\partial U}{\partial x} = m\ddot{x}$. As has been seen, that is the one where the action S is minimal (see Figure 1.1), such that any other position x must be rejected. Instead, the new non deterministic formulation of mechanics consists in admitting that ALL intermediate positions of the particle are in principle *possible*.

This admitted, it stands to reason that if the particle occupies position x_1 at time t_1, after a short time $\Delta t = t - t_1$ it has virtually occupied all space. In order to be more precise about this virtuality, we define the propagation amplitude $K(x,t) \equiv \langle x\,t \mid x_1\,t_1 \rangle$ as a COMPLEX number, that characterizes the motion from x_1 to x. Its *phase* makes possible the *interference* between propagation amplitudes through different intermediate positions x in the transition from x_1 to x_2. The *modulus* of this complex number only depends on the time interval $t - t_1$. The propagation amplitude for the transition from $(x_1\,t_1)$ to $(x_2\,t_2)$ satisfies the following postulate of propagation:

$$\langle x_2\,t_2 \mid x_1\,t_1 \rangle = \int_{-\infty}^{+\infty} \langle x_2\,t_2 \mid x\,t \rangle \cdot \langle x\,t \mid x_1\,t_1 \rangle \, dx, \qquad (3.1)$$

where t is an intermediate time $t \in (t_1, t_2)$. The integration (i.e. the *sum*) of this product of complex numbers is done over all the virtual space of intermediate positions x (real, of course). This expression defines the fundamental property that propagation amplitudes must fulfill [1]. It can be thought up as follows: "in order to move from one point to another, objects must probe all positions in space".

In a deterministic (classical) mechanics, such amplitude would simply be either 1 (the motion from $(x_1 t_1)$ to (x_2,t_2) is *possible*, for a given trajectory), or *zero* otherwise. In other words, the only intermediate point x that could contribute to the integral (3.1) at time t, would be the one meeting Newton's law. By contrast, in Quantum Mechanics the zero amplitude never happens, and it is possible to go from $+1$ to -1 through a continuum of values on the *unit circle* of the complex plane, making *interference* possible.

---

[1] the amplitude (3.1) arises from a sum (integral) over many different amplitudes (paths), which simultaneously contribute to the motion. This is often referred to in the literature as superposition principle. We shall see in Chapter 6 how this principle immediately extends to the state of motion itself.

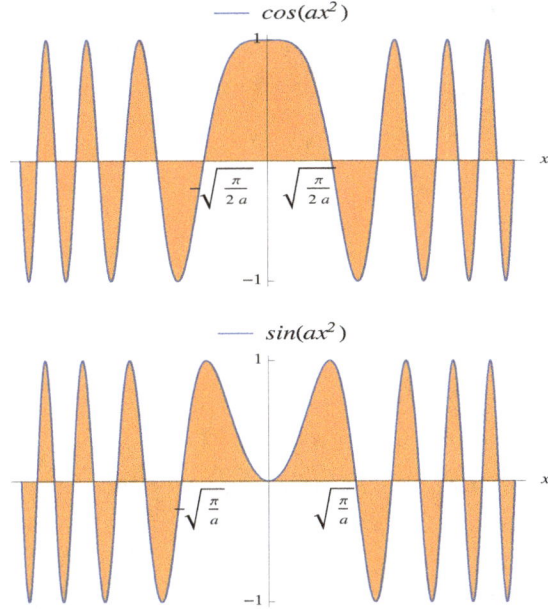

Figure 3.1: *Graphical representation of the Fresnel functions, which are the real (top) and imaginary (bottom) part of the function e^{iax^2}.*

A proper definition of the propagation amplitude, in a new formulation of physics, must achieve that, in the case of macroscopic motion (for long time intervals $\Delta t \gg \frac{h}{E}$), the mechanism of interference between different "jumps" x far away from the classical trajectory, must be strongly destructive in order to verify Newton's law with sufficient precision.

Postulate (3.1) is calling for the utilization of the *exponential function*, with its defining property $\exp(a+b) = \exp(a)\exp(b)$. Moreover, we are compelled to use an *oscillatory function* in order to achieve interference. Therefore, the use of complex numbers for the description of the laws of motion becomes *unavoidable* [2], under the form $e^{if(x)}$. Interestingly, these are not needed in the formulation of Classical Mechanics.

Such definition of the propagation amplitude for $\Delta t \to 0$, which we take here as a postulate, was given in 1948 by the American physicist Richard P. Feynman. This postulate is a realization of equation (3.1), and should be adopted together with it.

---

[2] we shall gain in Chapter 5 more insight into this point.

Chapter 3. THE FEYNMAN PROPAGATION

Feynman's propagation principle: *The propagation amplitude for the motion of a particle of mass m, subject to a potential $U(x,t)$, from point x_1 at time t_1 to point x at time t, in the limit $\Delta t = t - t_1 \to 0$, is given by:*

$$K(x,t;x_1 t_1) \equiv \langle x\, t \mid x_1\, t_1 \rangle = A\, e^{\frac{iS}{\hbar}} = A\, e^{\frac{i}{\hbar}L\Delta t}$$

$$= \sqrt{\frac{m}{i 2\pi\hbar\Delta t}} \exp\left[\frac{i}{\hbar}\left(\frac{m(x-x_1)^2}{2\Delta t} - U\left(\frac{x+x_1}{2},t\right)\Delta t\right)\right] \quad (3.2)$$

where $S = L\Delta t = (T - U)\Delta t$ is the full classical action that corresponds to the motion in the space-time interval from (x_1, t_1) to (x, t).

In order to deduce the value of the coefficient $A = \sqrt{m/(i2\pi\hbar\Delta t)}$, one should first realize that in the limit $\Delta t \to 0$ the potential energy has no influence on this factor, by simply comparing the opposite asymptotic behaviour of the two terms in the exponent, thus the exact value of A actually corresponds to the free motion case $(U(x,t) = 0)$ [3].

The coefficient A is obtained from postulate (3.1) after dividing the interval Δt into two halves $\Delta t/2$, and adding the exponents of the respective propagators. This is left as an exercise to the student, using for that purpose the value of the integral [4]

$$\int_{-\infty}^{+\infty} e^{ax^2+bx}\, dx = \sqrt{\frac{\pi}{-a}} e^{-b^2/4a} \quad a,b \in \mathbb{C} \quad \mathrm{Re}(a) \leq 0. \quad (3.3)$$

When making the above calculation with equation (3.1), note that the left-hand side is proportional to $A(\Delta t)$, whereas the right-hand side is proportional to $A^2(\Delta t/2)\sqrt{\Delta t}$. The solution is $A \propto 1/\sqrt{\Delta t}$, hence we have $A(\Delta t/2) = \sqrt{2} A(\Delta t)$, so that the desired result follows immediately.

In the case of free motion, the real and imaginary parts of the propagator are the well known Fresnel functions, which integral is obtained by setting $b = 0$ and $a = -ia$, $a \in \mathbb{R}$, in equation (3.3):

$$\int_{-\infty}^{+\infty} \cos(ax^2)\, dx = \int_{-\infty}^{+\infty} \sin(ax^2)\, dx = \sqrt{\frac{\pi}{2a}}\,.$$

---

[3] this is only valid for potentials having an asymptotic increase for $x \to \pm\infty$ at most quadratic, i.e. if $|U(x,t)| \leq C(x-x_1)^2$ $\forall x$, for some constant C.

[4] by making use of the integral (3.3) it can easily be shown that indeed Feynman's propagator satisfies postulate (3.1). Note that, by the same token, this postulate is also satisfied by a *real* propagator obtained from (3.2) after the replacement $i \to -1$. Such an option does not conform to Quantum Mechanics, lacking its crucial oscillatory behaviour from the exponent. It conforms instead to the diffusion processes governed by Brownian motion, and related phenomenology.

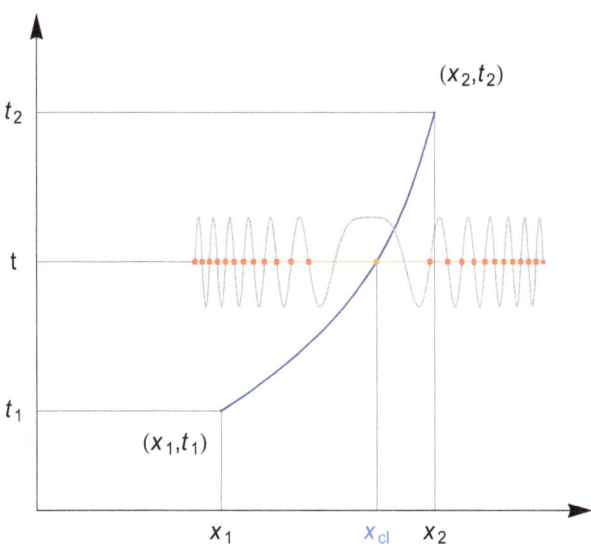

Figure 3.2: *Positions occupied by a moving body (orange line and red dots) at an intermediate time t when moving from x_1 to x_2 in a short time interval, far away from the classical trajectory (blue line), and the real part of the amplitude assigned to them by the Feynman propagator (grey line), on an arbitrary scale.*

A graphical representation of these functions is shown in Figure 3.1. Detailed observation of the curve reveals the reason why these integrals are convergent: due to the rapid oscillation of the phase for values $x \to \pm\infty$, the positive and negative contributions cancel out more precisely the larger the $|x|$ values, so that the main contribution to the integral comes from $|x|$ smaller than the first zeros [5] ($|x| \lesssim \sqrt{\frac{\pi}{a}}$).

Coming back to Feynman's principle, we grasp the physical importance of the above, since according to (3.1), all contributions from the intermediate points x, shown in Figure 3.2, must be summed up. The largest contribution comes from an interval around the point x_{cl} of the classical trajectory $(x_{cl} - \Delta x, x_{cl} + \Delta x)$ with $\Delta x \lesssim \sqrt{\frac{\pi \hbar \Delta t}{m}}$. Meaning that far-away separations from the classical trajectory contribute very little to the motion, due to the aforementioned cancellations. Fancifully, we might say that "the paths going through Beijing and through Shanghai cancel one another".

---

[5] the zeros are located at $x_n = \pm\sqrt{(2n-1)\pi/2a}$ for the cosine function and $x_n = \pm\sqrt{n\pi/2a}$ for the sine function, with $n = 1, 2, \ldots \infty$.

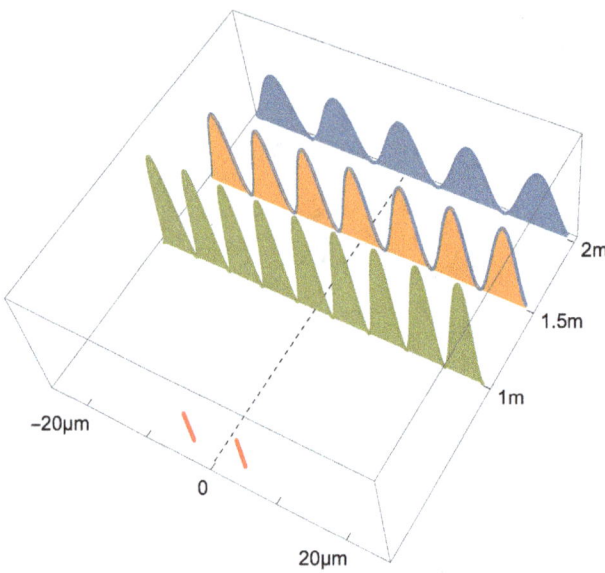

Figure 3.3: *Distribution of the probability density to find a Helium atom on the screen in a double slit experiment, according to the 1D Feynman propagator* (3.2) *with* $U(x,t) = 0$. *The slit gap is* $d = 8\,\mu m$. *A longer distance to the screen means increasing the visibility of the maxima (for a given pixel size), for their wider spacing. This is due to the longer time lag (t), for a given He atom velocity of* $v = p/m = 2.15\,kms^{-1}$ ($\lambda = h/p = 46\,pm$). *This value is well within its Maxwell distribution at room temperature, and is assumed constant throughout. The Helium atomic mass is* $4.00\,u$.

The phenomenon of *interference*, that is present in Quantum Mechanics through the complex expression (3.2) in 1D, is made the clearest when we restrict the initial propagation points (time t_1) to *only two*: x_1 and x_2, at a distance $d = x_2 - x_1$, as it is precisely done in the double slit experiments [6]. In these experiments, the particle of mass m proceeds simultaneously ($t_1 = 0$) in opposite directions, from *both* coordinates $x_1 = -d/2$ and $x_2 = +d/2$, interfering with itself to produce *visible* maxima and minima of the probability to locate the particle. As we shall see in Chapter 6, the probability density to find the particle at the point with coordinate x is defined by the squared modulus of the propagator through the expression: $|K(x,t;-\frac{d}{2}0) + K(x,t;\frac{d}{2}0)|^2 = C\left[1 + \cos\left(\frac{md}{\hbar t} \cdot x\right)\right] \in [0, 2C]$, with $t = \frac{mL}{p}$ being the propagation time to reach the screen at a distance L from the double slit. See Figure 3.3.

---

[6] in such devices, that are assessed in the course, the Feynman propagation technically takes place in 2D. Particularly revealing have been those performed with Helium atoms in 1997 (Ch. Kurtsiefer et al., Nature 386 (1997), 150-153), admirably described in the book "Quantum Physics" by J.S. Townsend (2010).

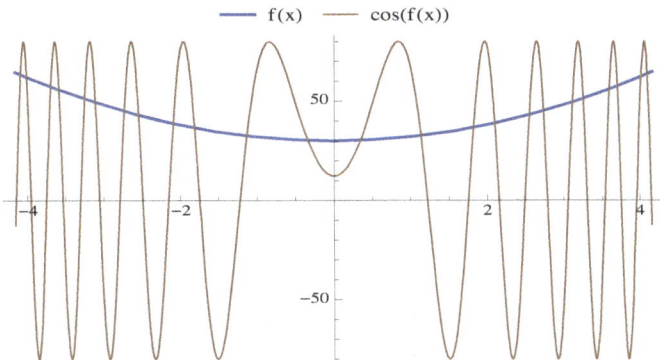

Figure 3.4: *The function cos(f(x)) shows less oscillations near the minimum of f(x).*

Note the interference actually takes place in 1D (*x*-direction), while it gets amplified as the particle travels to the screen with velocity $v = p/m$ in the direction perpendicular to the double slit. We leave as a simple exercise to prove the above (real positive) expression for the probability density.

More detail will be seen in Chapter 4 about the significance of Figure 3.2, and its consequences. Section 3.1 next is not essential to comprehend the rest of this course, and may be skipped.

3.1 Exact propagation over a finite time

Expression (3.2) is valid for an infinitesimal time interval Δt. If a finite time interval is required, then it is necessary to apply (3.2) repeatedly, at successive time steps dt, taking into account that, at each step, the particle may move from any previous point to any other point in space.

Only for the student specifically interested, we indicate below the detailed way in which such integration is performed.

The interval $\Delta t = t_b - t_a$ is divided into small steps $\varepsilon = t_{i+1} - t_i$ with $\Delta t = N\varepsilon$, so that at each time t_i we select some arbitrary point x_i and construct a path by connecting all points so selected (x_i, t_i), for $i = 0, \ldots N$, with $t_0 = t_a$ and $t_N = t_b$. We evaluate the Lagrangian $L(x, \dot{x}, t)$ at each point (x_i, t_i) and then apply the propagator (3.2) at each step

$$K(i+1, i) = A \exp\left[\frac{i\varepsilon}{\hbar} L\left(\frac{x_{i+1} + x_i}{2}, \frac{x_{i+1} - x_i}{\varepsilon}, \frac{t_{i+1} + t_i}{2}\right)\right],$$

Thus the propagator over the finite interval is just the product of all of them

$$K(x_b,t_b;x_a,t_a) = \lim_{\varepsilon \to 0} \prod_{i=0}^{N-1} K(i+1,i) , \qquad (3.4)$$

which corresponds to the detailed expression

$$K(x_b,t_b;x_a,t_a) = \lim_{\varepsilon \to 0} A \int \int \cdots \int e^{\frac{i}{\hbar}S[b,a]} (Adx_1)(Adx_2)\ldots(Adx_{N-1}) \qquad (3.5)$$

where $S[b,a] = \int_{t_a}^{t_b} L(x,\dot{x},t)dt$ is the line integral obtained from the path of straight-line steps (x_i,t_i) described above and A is the factor calculated earlier in this chapter from equation (3.3). The integration over the spatial coordinates is performed at each time t_i according to the postulate of propagation (3.1), which gives rise to a $(N-1)$–dimensional multiple integral. Finally the limit $N \to \infty$ is taken (that is equivalent to $\varepsilon \to 0$).

Expression (3.5) is known as Feynman's path integral in the literature. As said above, it is a $(N-1)$–dimensional integral over the space coordinates, where the time sum comes to complete the action integral in the exponent, in the limit $\varepsilon \to 0$. It is customary to use the succint notation

$$K(b,a) = \int_a^b e^{\frac{i}{\hbar}S[b,a]} \mathscr{D}x(t) , \qquad (3.6)$$

where the symbol $\mathscr{D}x(t)$ reminds us that the multiple spatial integration (3.5) is equivalent in fact to a sum over *all possible trajectories* $x(t)$ starting from the point $x_a = (x_0,t_0)$ to reach the point $x_b = (x_N,t_N)$ of space-time.

Be noted that, even if present in the above sum, *differentiable* trajectories only constitute a tiny fraction of the immensity of the non differentiable trajectories that build the motion.

For the discussion that follows, particularly the derivation of Schrödinger's equation in Chapter 5, the consideration of the infinitesimal time interval in Eq. (3.2) will suffice.

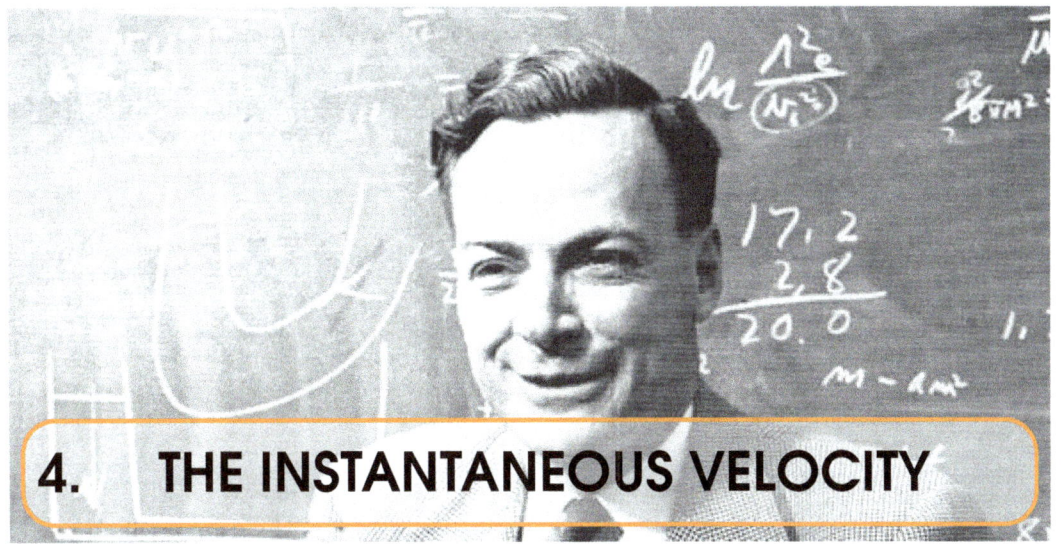

4. THE INSTANTANEOUS VELOCITY

Let us see below how the intermediate position $x_1 < x < x_2$ that verifies Newton's second law is, according to Feynman's hypothesis, precisely the central value around which the particle's position fluctuates. Consider again the sequence represented in Figure 3.2, where a particle of mass m moves from (x_1, t_1) to (x_2, t_2) through the intermediate position x (at time $t = (t_1 + t_2)/2$), with $\Delta t = t_2 - t_1$. Then the amplitude for the transition $1 \to 2$, according to (3.1) is given by:

$$\langle x_2\, t_2 \mid x_1\, t_1 \rangle = \int_{-\infty}^{+\infty} \langle x_2\, t_2 \mid x\, t \rangle \langle x\, t \mid x_1\, t_1 \rangle\, dx,$$

which, as stated in (3.2), can be expressed as:

$$\int_{-\infty}^{+\infty} A^2\, e^{\frac{i}{\hbar}\left[\left(\frac{m}{2}\frac{(x-x_1)^2}{(\Delta t/2)^2} - U\left(\frac{x_1+x}{2}\right)\right)\frac{\Delta t}{2}\right]}\, e^{\frac{i}{\hbar}\left[\left(\frac{m}{2}\frac{(x_2-x)^2}{(\Delta t/2)^2} - U\left(\frac{x+x_2}{2}\right)\right)\frac{\Delta t}{2}\right]} dx$$

$$= \int_{-\infty}^{+\infty} A^2\, e^{\frac{i}{\hbar}\frac{\Delta t}{2}\left[\frac{m}{2}\frac{(x-x_1)^2+(x_2-x)^2}{(\Delta t/2)^2} - \left(U\left(\frac{x_1+x}{2}\right)+U\left(\frac{x+x_2}{2}\right)\right)\right]} dx$$

As earlier indicated, the main contribution to the integral (real and imaginary parts) comes from the region where x minimizes the *phase* (the expression under square brackets). This is the so-called *stationary phase* condition, in the literature.

In general, for a function $f(x)$ having a minimum at $x = x_c$, the value of the integral $\int_{-\infty}^{+\infty} \cos(f(x))\, dx$, which is the real part of the above expression, gets the largest contribution from those values $x \simeq x_c$ where the cosine has the least number of oscillations per unit length, as illustrated in Figure 3.4.

Yet the value of x where the expression in square brackets is minimal is precisely that verifying Newton's second law ($-\frac{\partial U}{\partial x} = m\ddot{x}$), as was shown in Chapter 1, upon differentiation of equation (1.4).

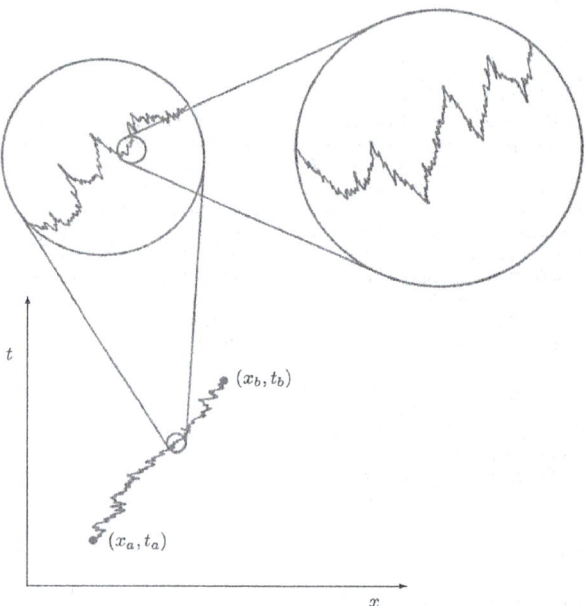

Figure 4.1: Irregular path followed by a particle when looked in detail in a space-time diagram: the trajectory is not differentiable. Original drawing by Richard P. Feynman in Quantum Mechanics and Path Integrals, 1948.

Thus we see that the effect of destructive interference derived from expression (3.2) is essential to produce an amplitude suppression for positions far away from the classical trajectory of the particle, during the observation time Δt. Let us recall that the region where these fluctuations become important is determined by $\Delta x \lesssim \sqrt{\pi \hbar \Delta t / m}$, showing that even if the spot Δx "blurred" by fluctuations becomes infinitely small in the limit $\Delta t \to 0$, it does not do so *linearly*, but proportional to $\sqrt{\Delta t}$ (less rapidly). Hence the ratio $\Delta x/\Delta t$ is not finite, in the above limit, but diverges as $1/\sqrt{\Delta t}$, which tells us that the instantaneous velocity *does not make sense*. Its absolute value becomes infinite, if we take the non relativistic expression for the kinetic energy [1].

---

[1] we recall here the relativistic consideration made in Section 2.1, which takes the velocity to c.

This important conclusion, which is ultimately a consequence of the discrete (and *non zero*) character of the reduced action, undoubtedly breaks up with preconceived ideas about the differentiability of trajectories. It is clear that particle observation during shorter and shorter time intervals will produce higher and higher velocities.

On the other hand, if the observation time interval is sufficiently long, as for instance, when a photography is taken with exposure time some fraction of a second (10^{-2}–$10^{-3}s$), then the motion appears to be perfectly continuous, without fluctuations. This idea can be appreciated in Figure 4.1, taken from the book by Feynman and Hibbs *Quantum Mechanics and Path Integrals*, Dover (2010).

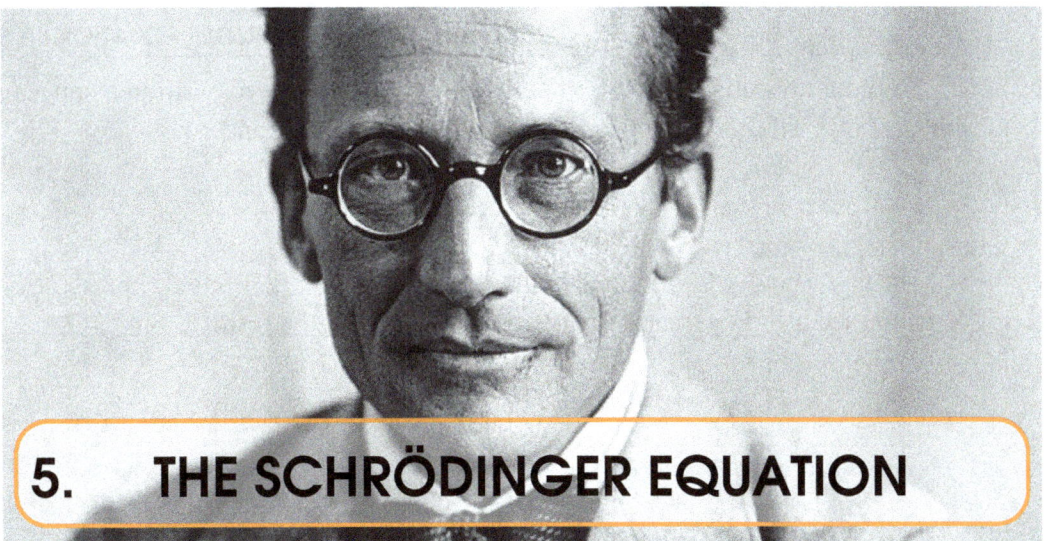

5. THE SCHRÖDINGER EQUATION

We have seen up to now how the particle's motion in the limit $\Delta t \to 0$ can be represented by means of a space-time propagation amplitude $\langle x\,t \mid x_1\,t_1\rangle$, which is actually a complex function of the real variables (x,t). This idea has allowed us to reconcile the discontinuous character of the action at short time scales with a mathematical description that renders the use of differentiable functions possible, for the study of motion.

Despite its conceptual riches, the practical use of expression (3.6) to propagate the particle from (x_1,t_1) to (x_2,t_2) when $\Delta t = t_2 - t_1$ is *finite* (not infinitesimal), requires the introduction of new mathematical tools of integration, which we shall not develop here. Instead, we shall show an easier way to use expression (3.6), based on partial differential equations. All it takes is to realize that the function $K(x,t) \equiv \langle x\,t \mid x_1\,t_1\rangle$ actually behaves as a strongly dispersive wave in the coordinates (x,t).

Indeed, let us show that, according to expression (3.1), and to the form (3.2) of the propagation amplitude $K(x,t) \equiv \langle x\,t \mid x_1\,t_1\rangle$ for $\Delta t = t - t_1 \to 0$, the function $K(x,t)$ satisfies the following partial differential equation, called Schrödinger's equation:

$$i\hbar\frac{\partial K(x,t)}{\partial t} = \frac{-\hbar^2}{2m}\frac{\partial^2 K(x,t)}{\partial x^2} + U(x,t)K(x,t)\,. \tag{5.1}$$

For this purpose, let us take into account that the point $(x, t+\Delta t)$ is reached from all space points $x - \xi$ at an earlier time t.

Chapter 5. THE SCHRÖDINGER EQUATION

So the amplitude $K(x, t+\Delta t) \equiv \langle x\, t+\Delta t \mid x_1\, t_1 \rangle$ can be written as an integral over all these points $x - \xi$, according to expression (3.1)

$$K(x, t+\Delta t) = \int_{-\infty}^{+\infty} K(x, t+\Delta t\, ;\, x-\xi, t) K(x-\xi, t) d\xi\,.$$

The first factor in the integrand $K(x, t+\Delta t\, ;\, x-\xi, t) \equiv \langle x, t+\Delta t \mid x-\xi, t\rangle$ is expressed as a Feynman propagator, thus yielding the result

$$K(x, t+\Delta t) = \sqrt{\frac{m}{2\pi i\hbar \Delta t}} \int_{-\infty}^{+\infty} \exp\left(\frac{i}{\hbar}\frac{m\xi^2}{2\Delta t}\right) \exp\left(\frac{i}{\hbar}[-U(x-\xi/2, t)\Delta t]\right) K(x-\xi, t) d\xi \quad (5.2)$$

Given that the function $K(x,t)$ is infinitely differentiable at the point (x,t) with $t \neq t_1$, in order to relate its partial derivatives we write down its expansion in powers of Δt

$$K(x, t+\Delta t) = K(x, t) + \Delta t \frac{\partial}{\partial t} K(x, t) + \ldots\,,$$

and in powers of ξ

$$K(x-\xi, t) = K(x, t) - \xi \frac{\partial K(x, t)}{\partial x} + \frac{\xi^2}{2} \frac{\partial^2 K(x, t)}{\partial x^2} + \ldots\,, \quad (5.3)$$

as well as the expansion of the exponential function

$$e^{-\frac{i}{\hbar}U(x-\xi/2, t)\Delta t} = 1 - \frac{i}{\hbar} U(x-\xi/2, t)\Delta t + \cdots = 1 - \frac{i}{\hbar} U(x,t)\Delta t + \frac{i}{\hbar}\frac{\xi}{2}\frac{\partial U}{\partial x}\cdot\Delta t - \frac{i}{\hbar}\frac{\xi^2}{4}\frac{\partial^2 U}{\partial x^2}\cdot\Delta t + \ldots \quad (5.4)$$

It is clear that when making the crossed products between the expansions (5.3) and (5.4) on the right-hand side of equation (5.2), 12 terms will appear, of which those with odd powers of ξ vanish out, after integration from $-\infty$ and $+\infty$. In order to evaluate the even powers, we use the following integrals [1].

$$\int_{-\infty}^{+\infty} e^{\frac{i}{\hbar}\frac{m\xi^2}{2\Delta t}} d\xi = \sqrt{\frac{2\pi i\hbar \Delta t}{m}}$$

and

$$\int_{-\infty}^{+\infty} \xi^2 e^{\frac{i}{\hbar}\frac{m\xi^2}{2\Delta t}} d\xi = \sqrt{2\pi}\left(\frac{i\hbar \Delta t}{m}\right)^{3/2}$$

---

[1] the first one is a particular case of (3.3), and the second one is obtained from the first by differentiation with respect to the coefficient multipying ξ^2

Thus the remaining terms on the right-hand side of (5.2) are proportional to powers of Δt, of which we neglect $(\Delta t)^2$, $(\Delta t)^3$, $\cdots$ in the limit $\Delta t \to 0$.

The expression finally obtained is

$$\Delta t \frac{\partial K(x,t)}{\partial t} = \sqrt{\frac{m}{2\pi i\hbar \Delta t}} \sqrt{2\pi} \left(\frac{i\hbar \Delta t}{m}\right)^{3/2} \frac{1}{2}\frac{\partial^2 K(x,t)}{\partial x^2} - \frac{i}{\hbar}\Delta t U(x,t)K(x,t),$$

from which equation (5.1) follows after elimination of Δt from both sides. Note that terms containing $\frac{\partial^2 U}{\partial x^2}K$, $\frac{\partial U}{\partial x}\frac{\partial K}{\partial x}$, and $U\frac{\partial^2 K}{\partial x^2}$, come out proportional to $(\Delta t)^2$ and do not contribute in the limit $\Delta t \to 0$.

The differential equation (5.1) is a fundamental tool for all applications of Quantum Mechanics. It was discovered in 1926 by the Austrian physicist Erwin Schrödinger.

This differential equation is *complex* as a direct consequence of the presence of the imaginary unit i in Feynman's propagator. We could then wonder whether adopting the *real* propagator given by the expression: $K_c(x - x_1, \Delta t) = \sqrt{\alpha/\pi} \cos[\alpha(x - x_1)^2 - U(\frac{x+x_1}{2}, t)\Delta t/\hbar + \beta]$, $\alpha \equiv \frac{m}{2\hbar\Delta t}$, that is nothing but the real part of Feynman's propagator with a suitable phase β, would have given rise to a *real* differential equation that could replace Schrödinger's equation.

In order to answer the question raised, the first thing to do is figure out whether or not the real propagator K_c verifies the postulate or propagation (3.1), for $\Delta t \to 0$. Surprisingly, the response is positive for the *free* motion ($U(x,t) = 0$), subject to the condition $\beta = -\pi/4$. However, the above postulate is no longer verified for the general case with $U(x,t) \neq C$ [2].

This corroborates the *necessity* of complex numbers, in order to achieve a viable representation of Feynman's postulate (3.1) based on assigning a *phase* to the ratio $S/\hbar$ in (3.2), the cosine function not being able to generate any partial differential equation representing Feynman's propagation.

---

[2] to better understand this, take into account the expression $\cos(A+B) = 2\cos A \cos B - \cos(A-B)$, where a *second term* appears with a *non zero* integral over $(-\infty, +\infty)$, in the presence of a potential $U(x,t) \neq 0$.

6.   THE WAVE FUNCTION

We have defined $K(x_2,t_2;x_1,t_1)$ as the propagation amplitude for a particle of mass m to move from (x_1,t_1) to (x_2,t_2) under a potential $U(x,t)$, and we have done it using the classical action. Contrary to what happens in Classical Mechanics, where the motion may or may not be possible for a given trajectory, in Quantum Mechanics the motion is always possible, since the exponential function is never zero. Thus the particle initially located at point x_1, will have virtually propagated to all space points after a very small time $t_2 - t_1 > 0$ and each point is in turn subject to subsequent propagation. Hence it becomes necessary to define the *state of occupation of the space* that a particle has at a given time.

It is of interest to consider the propagation amplitude for a particle to arrive at a given point, without any particular information about its previous motion. Following Feynman, a complex function $\psi(x,t)$ can then be defined as the `total amplitude` to reach the point (x,t). This amplitude is called the wave function. There is no conceptual difference with respect to the propagation amplitude we have seen. In fact, the propagator $K(x,t;x_1,t_1)$ is a wave function itself, since it represents a particular amplitude to reach the space-time point (x,t), specifically from (x_1,t_1). When we use the notation of the wave function, it means we are not interested in the particle's prior motion.

As $\psi(x,t)$ is a propagation amplitude, it complies with the general postulate of propagation (3.1).

Given that the postulate is valid for *all* points x_1, the wave function must satisfy the integral equation

$$\psi(x,t) = \int_{-\infty}^{+\infty} K(x,t;x_1,t_1)\, \psi(x_1,t_1)dx_1 \ . \tag{6.1}$$

This result can be stated in physical terms: the total amplitude to reach (x,t) is the sum (integral) over all possible values of x_1, of the total amplitude to reach the point (x_1,t_1) ($\psi(x_1,t_1)$), times the amplitude to move from x_1 to x ($K(x,t;x_1,t_1)$). The effects of the past history of the particle can then be expressed in terms of a single function. Equation (6.1) holds with the exact form of the propagator given in (3.5), and its approximation by using expression (3.2) for the propagator, for short time intervals, is particularly useful.

Just as the propagator satisfies the Schrödinger equation

$$i\hbar \frac{\partial K(x,t)}{\partial t} = \frac{-\hbar^2}{2m} \frac{\partial^2 K(x,t)}{\partial x^2} + U(x,t) K(x,t) \ ,$$

so does the wave function $\psi(x,t)$

$$i\hbar \frac{\partial \psi}{\partial t} = \left(\frac{-\hbar^2}{2m} \frac{\partial^2}{\partial x^2} + U(x,t) \right) \psi \ . \tag{6.2}$$

Indeed the above is easily shown by differentiating under the integral sign in (6.1), the details of which we leave as an exercise to the student. The full mathematical content of equation (6.1) can now be understood, given that the function $\psi(x_1,t_1)$ plays the role of a *single* (arbitrary) initial condition for the time evolution of $\psi(x,t)$ [1].

Thus the complex function $\psi(x,t)$ is taken as a definition of the *state of motion* of the particle at time t. Strictly associated with it comes its probabilistic interpretation, introduced by the German physicist Max Born in 1926: the probability density to find the particle at $(x,x+dx)$, as a result of a measurement, is given by

$$\frac{dP(x,t)}{dx} = |\psi(x,t)|^2 \ .$$

---

[1] note that Schrödinger's equation is first order in the time derivative ($\partial/\partial t$), so its solutions depend on a single integration constant. Therefore it is not required to specify the function $\frac{\partial \psi}{\partial t}$ at $t=0$, as initial condition. This is in contrast to Newton's equation, and to the wave equation, both of which are second order in t, containing $\frac{\partial^2}{\partial t^2}$.

A consequence of the above is the normalization condition, stating that the total probability for the particle to be somewhere in space must be one:

$$\int_{-\infty}^{+\infty} |\psi(x,t)|^2 dx = 1 \ . \tag{6.3}$$

Note the implicit requirement that $\psi(x)$ has to be square integrable: $\int_{-\infty}^{+\infty} |\psi(x,t)|^2 dx < +\infty$, at all times t.

An important property of the wave function is that its multiplication by any global phase factor $e^{i\theta}$, with $\theta \in [0, 2\pi]$ independent of space and time coordinates, does not change any of its physical properties, thus representing exactly the same state. This is a natural consequence of the undefined zero level of the potential energy in Classical Mechanics, and of the way it acts in Feynman's propagator (3.2). Indeed, a redefinition $U(x,t) \to U(x,t) + C$ turns out to be indistinguishable from a change of the origin ($t = 0$) of *time* $t \to t + t_0$.

It should be made clear that the non physical character of the global phase of ψ only refers to an *isolated* particle, that is, for the *complete* wave function, not when it is part of a linear combination with others.

Hence a 1–1 correspondance can be established between the states of motion of a particle and the *complex functions* that verify equations (6.2) and (6.3), except for an arbitrary phase factor $e^{i\theta}$. States are often denoted by the Dirac symbol $|\psi\rangle$.

7. DE BROGLIE WAVE AND FOURIER TRANSFORMATION

The Schrödinger equation (6.2) is a partial second order differential equation that involves $\partial/\partial t$, pertaining to a group also formed by the wave equation (having $\partial^2/\partial t^2$, that describes non-dispersive waves) and the diffusion equation, that results from replacing $i \to -1$, and describes phenomena related to Brownian motion. Let us consider as a possible solution of (6.2) for free motion ($U(x,t) = 0$) a monochromatic plane wave in 1D

$$\psi(x,t) = A e^{i(kx-\omega t)} . \tag{7.1}$$

We know this expression represents a propagating wave along the positive X axis, with angular frequency ω, wave number k and velocity $v_p = \omega/k$, that fulfils the wave equation $\partial^2\psi/\partial x^2 - (1/v_p^2)\partial^2\psi/\partial t^2 = 0$, hence it will not meet equation (6.2) with $U(x,t) = 0$, unless the dispersion relation $\omega = \omega(k)$ is changed, to take the precise form

$$\omega(k) = \frac{\hbar k^2}{2m} , \tag{7.2}$$

as it is easily shown, by applying equation (6.2) to (7.1).

We shall associate the plane wave represented by the solution (7.1) of Schrödinger's equation with a particular state of the particle, following the historical path laid down by the French physicist Louis De Broglie, who conjectured in 1923 the following idea:

De Broglie's hypothesis

Every moving body with momentum p carries with it a wave, which is inherent with its state of motion, with wavelength

$$\lambda = \frac{h}{p}, \tag{7.3}$$

where h is the Planck constant.

Indeed this wave, represented in Figure 7.1 for $t = 0$, with the dispersion relation (7.2), is nothing but the state $\psi(x,t)$ of a particle in free motion with momentum $p = \hbar k$, if we associate its velocity with the *group velocity* $v_g = \frac{d\omega}{dk}$ of the wave[1], since

$$p = mv_g = m\frac{d\omega}{dk} = m\frac{\hbar k}{m} = \hbar k = \frac{h}{\lambda},$$

the kinetic energy of the particle being associated with its frequency ω:

$$E = \frac{p^2}{2m} = \frac{\hbar^2 k^2}{2m} = \hbar\omega.$$

Formulae $p = \hbar k$, and $E = \hbar\omega$, are known as **De Broglie's relations** in the literature. If the relativistic dispersion relation $\omega = \sqrt{k^2c^2 + (mc/\hbar)^2}$ is used, associated with the total energy $E = \sqrt{p^2c^2 + m^2c^4}$, then the plane wave (7.1) retains its physical meaning, in a perfectly relativistic way. However the Schrödinger equation is no longer satisfied, and it needs to be replaced by a relativistic one (Klein-Gordon or Dirac, not dealt with here).

The strongly dispersive character becomes evident when we add waves with different values of λ to form wave packets, since the propagation velocities of their phases are inversely proportional to their wavelengths ($v_p = \omega/k = h/(2m\lambda)$), which causes the wave packets lose their shape.

With the momentum and energy assignment granted by De Broglie's relations, the phase of (7.1) exactly coincides with the *classical action S* divided by $\hbar$, and therefore it also represents the *propagation amplitude* for a *free* particle moving with constant velocity across the space, according to Feynman's propagator. Indeed, $x = vt$ and $L = E = T = mv^2/2$, thus

$$kx - \omega t = \frac{px}{\hbar} - \frac{Et}{\hbar} = \frac{1}{\hbar}(2\frac{mv^2}{2}t - Et) = \frac{1}{\hbar}Lt = \frac{1}{\hbar}S.$$

---

[1] in fact De Broglie's original *relativistic* reasoning to introduce such wave was quite different from the above. His wave was postulated to be in phase with what he deemed an *internal periodic process* associated to the energy mc^2. The *dispersive character* of the wave, through a relation $\omega(k) \neq vk$, allowed to solve the conspicuous conflict between the frequency increase due to a mass increasing with velocity ($\hbar\omega = mc^2\gamma$), and the frequency decrease resulting from Lorentz's time dilation ($\hbar\omega_1 = mc^2/\gamma$).

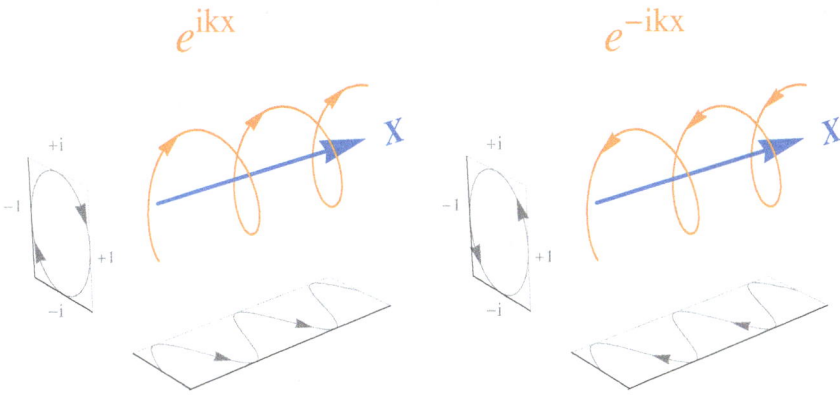

Figure 7.1: *De Broglie's plane wave representing a particle moving along the X axis, for both signs of the wave number ±k. Note how the function value is never zero, showing a dextro/levo character associated with forward/backward motion.*

At all times t, the solution (7.1) shows a constant probability density across the space

$$|\psi(x,t)|^2 = |A|^2 = \text{constant} \quad \forall x \in \mathbb{R}. \tag{7.4}$$

Which means a *mathematical idealization*, since, in general, we do not have wave packets in physics that are totally monochromatic, but instead they keep a constant wavelength λ over a certain maximum length (called *coherence length* in the literature), and decay to zero outside. A consequence of the ideal character of the *plane wave* states is that the correct value of the constant A in (7.1) cannot be simply determined from the normalization condition (6.3), since this integral diverges. In other words, we are not talking about a square-integrable function.

At a given time we define the Fourier transform of any wave function $\psi(x)$ as

$$f(k) \equiv \frac{1}{\sqrt{2\pi}} \int_{-\infty}^{+\infty} \psi(x) \, e^{-ikx} dx. \tag{7.5}$$

The Fourier transformation is one of the most powerful mathematical tools ever invented, its properties and related theorems being found in many textbooks [2].

---

[2] see for instance "The Fourier Transform and its Aplications", R. N. Bracewell (2000).

The *inversion theorem* of the Fourier transformation states that the wave function can always be retrieved as

$$\psi(x) = \frac{1}{\sqrt{2\pi}} \int_{-\infty}^{+\infty} f(k)\, e^{ikx} dk , \qquad (7.6)$$

with the additional property that $\int_{-\infty}^{+\infty} |f(k)|^2 dk = 1$ [3].

This actually means that we can use plane waves e^{ikx} with different wavelengths to build any complex function defined on $\mathbb{R}$. Since $|\psi(x)|^2 dx$ represents the probability for a particle to be detected in $(x, x+dx)$, we should associate $|f(k)|^2 dk$ with the probability for the particle to be detected with momentum in $(p, p+dp) = \hbar(k, k+dk)$. Note that while x is measured in length units (m), k is measured in inverse length units (m^{-1}), or number of waves per unit length.

Just as the moving body simultaneously occupies a distributed region in the coordinate space, we must admit its velocity is not unique, the velocity space being also occupied in a continuous manner, according to the Fourier transform of $\psi(x)$. It should be emphasized that the complex functions $\psi(x)$ and $f(k)$ provide two equivalent descriptions of the same state of motion, since they contain exactly the same information. Such correlation between position and velocity of a moving body is an *unknown phenomenon* to Classical Mechanics.

---

[3] note the equality between the total integrals of $|\psi(x)|^2$ and $|f(k)|^2$ requires the presence of the factor $1/\sqrt{2\pi}$ in the definition of the Fourier transformation (7.5), which is specific to all its applications in Quantum Mechanics.

8. MEAN VALUES AND UNCERTAINTY

Since the particle's wave function stretches over all space, it is essential managing to calculate the mean value $\langle x \rangle$ of the coordinates of uniformly distributed pixels giving a signal in measurements of its position, and the dispersion Δx of the measurements around the mean value, i.e. the spatial extent over which the particle fluctuates. Given that $|\psi(x)|^2$ is a probability density function, these quantities are calculated as follows:

$$\langle x \rangle = \int_{-\infty}^{+\infty} x|\psi(x)|^2 dx = \int_{-\infty}^{+\infty} \psi^*(x)\,(x\psi(x))\,dx, \quad \text{and} \quad (8.1)$$

$$(\Delta x)^2 = \langle x^2 \rangle - \langle x \rangle^2,$$

where $\langle x^2 \rangle$ is the mean value of x^2, that is obviously calculated as

$$\langle x^2 \rangle = \int_{-\infty}^{+\infty} x^2 |\psi(x)|^2 dx.$$

Not being possible to define an instantaneous velocity, as has been described, it makes perfect sense after all to calculate the *mean velocity* $\langle v \rangle$, and the mean momentum $\langle p \rangle$ of the particle. The latter is determined, as discussed in the previous chapter, from the mean value of k in the Fourier transform

$$\langle p \rangle = \hbar \langle k \rangle = \hbar \int_{-\infty}^{+\infty} k|f(k)|^2 dk = m\langle v \rangle.$$

Yet it is possible to perform a more direct calculation of the mean momentum, without prior calculation of the Fourier transform, just by performing a single integral. In order to achieve that, we must learn some properties of the Fourier transformation.

In the ensemble of wave functions (mathematically the Hilbert space of square-integrable functions $L^2(\mathbb{R})$), a scalar product can be defined as

$$\langle \psi_1 | \psi_2 \rangle \equiv \int_{-\infty}^{+\infty} \psi_1(x)^* \psi_2(x) dx ,$$

where $\langle \psi_1 | \psi_2 \rangle$ is a *complex* number. Note that $\langle \psi_2 | \psi_1 \rangle = \langle \psi_1 | \psi_2 \rangle^*$.

It can easily be shown that the existence of such scalar product grants the Hilbert space the structure of a vector space. The Fourier transform then fulfils the following property, known in the literature as *generalized Parseval's identity*:

$$\int_{-\infty}^{+\infty} \psi_1(x)^* \psi_2(x) dx = \int_{-\infty}^{+\infty} f_1(k)^* f_2(k) dk \quad \forall \psi_{1,2} \in L^2(\mathbb{R}) ,$$

that is, the scalar product remains invariant under the transformation of each factor: $\langle \psi_1 | \psi_2 \rangle = \langle f_1 | f_2 \rangle$. In other words, it truly represents the projection of a quantum state into another, and the result is the same, whether it is performed in the position representation, or in the momenum representation of the wave function.

The above result allows to calculate $\langle p \rangle$ directly. Indeed, differentiation with respect to x on both sides of equation (7.6) renders

$$\frac{\partial \psi}{\partial x} = \frac{1}{\sqrt{2\pi}} \int_{-\infty}^{+\infty} ik f(k) e^{ikx} dk ,$$

which is telling us a general property, namely that the Fourier transform of the derivative $\frac{\partial \psi}{\partial x}$ is simply $ikf(k)$. Now, using the invariance of the scalar product, we get

$$\langle k \rangle = \int_{-\infty}^{+\infty} k |f(k)|^2 dk = \int_{-\infty}^{+\infty} \frac{1}{i} f(k)^* (ikf(k)) dk = \int_{-\infty}^{+\infty} \frac{1}{i} \psi^*(x) \frac{\partial \psi}{\partial x} dx ,$$

that, according to De Broglie's relation, amounts to

$$\langle p \rangle = \int_{-\infty}^{+\infty} \psi^*(x) (-i\hbar \frac{\partial \psi}{\partial x}) dx ,$$

which is the direct integral we were looking for.

The attentive reader will note that this expression shares a common mathematical structure with equation (8.1).

Indeed, both of them can be considered as particular cases of a more general definition of the mean value of a linear operator that represents any measurable physical quantity A:

$$\langle A \rangle = \int_{-\infty}^{+\infty} \psi^*(x)(A\psi)dx = \langle \psi | A\psi \rangle ,\qquad(8.2)$$

with $A = x$ (position) or p (momentum). In fact, a unique linear operator form can be found for every measurable magnitude A. Generally these can be obtained as a function of the above two.

Such operators are only intended to calculate mean values, and are *linear* mathematical applications mapping each element of the Hilbert space $L^2(\mathbb{R})$ into another element of the same space in a transformation ($\psi \to A\psi \in L^2(\mathbb{R})$). Note the linear character is rooted in the *superposition principle* we saw in Chapter 3.

As we have just seen, the momentum operator is represented by a partial derivative with respect to the coordinate of motion

$$p = -i\hbar \frac{\partial}{\partial x} .$$

Of course, the mean values $\langle A \rangle$ must be *real* for every wave function ψ, as the laboratory measurements are, for any physical magnitude A. This forces all physical operators A to be self-adjoint, i.e. for every pair of states $\psi_1, \psi_2 \in L^2(\mathbb{R})$: $\langle \psi_1 | A\psi_2 \rangle = \langle A\psi_1 | \psi_2 \rangle \equiv \langle \psi_1 | A | \psi_2 \rangle$ should be fulfilled, so that *real* $\langle A \rangle$ can be accomplished for $\psi = \psi_1 = \psi_2$, since $\langle \psi | A\psi \rangle = \langle A\psi | \psi \rangle^*$. The evaluation of $\langle A^2 \rangle$ also makes perfect sense. In this case, the action of A^2 should be understood as the repeated application $A(A\psi)$. Note the *profound analogy* of these operators with the *complex Hermitian matrices*, which are also linear self-adjoint operators on a vector space, of finite dimension. Similarly higher powers can be defined, power series, etc.

Individual laboratory measurements of the magnitude A, obtained from the same initial state ψ, will manifest the *non deterministic* character of Quantum Mechanics by throwing random values. However, we can predict with certitude the dispersion of the measurements ΔA around their mean value, through the expression

$$(\Delta A)^2 = \langle A^2 \rangle - \langle A \rangle^2 .\qquad(8.3)$$

If we want to calculate, for instance, the momentum dispersion Δp, $\langle p^2 \rangle$ can easily be obtained from the integral

$$\langle p^2 \rangle = \int_{-\infty}^{+\infty} \psi^*(x)\left(-\hbar^2 \frac{\partial^2 \psi}{\partial x^2}\right) dx \, .$$

The above integral is nonnegative for square-integrable functions that decay to zero at infinity, as can be shown using the integration by parts[1]. Thus we see that the operator $H = -\frac{\hbar^2}{2m}\frac{\partial^2}{\partial x^2} + U(x,t)$, which appeared in Schrödinger's equation (6.2), actually represents the total energy, since the first term (including the minus sign) represents the kinetic energy.

As a consequence of all the above, we now have a rule that allows us to evaluate, just by performing integrals, the mean value and the dispersion, not only of the position and the momentum of a particle, but of every magnitude A constructed from them. In other words, we have learned how to decode the information residing in the wave function $\psi(x)$, in order to make statistical predictions about the results of laboratory measurements performed on a particle.

As for matrices, the action of the linear operators A is not in general commutative. It is a simple exercise, for instance, to show that, for every wave function ψ, the following equation is verified

$$[x,p] \equiv xp - px = i\hbar \, ,$$

where it is to be warned about a usual habit in Quantum Mechanics, that should not lead to confusion: the same symbols can be used to designate the *values* of the physical magnitudes and the *operators* representing them. What is actually meant in the above equation is that: $(xp - px)\psi = i\hbar\psi$, $\forall \psi$.

---

[1] just take into account $-\int_{-\infty}^{\infty} \psi^* \frac{\partial^2 \psi}{\partial x^2} dx = -\left|\psi^*(x)\psi(x)\right|_{-\infty}^{+\infty} + \int_{-\infty}^{+\infty} \frac{\partial \psi^*}{\partial x}\frac{\partial \psi}{\partial x} dx = \int_{-\infty}^{+\infty} \left|\frac{\partial \psi(x)}{\partial x}\right|^2 dx \geq 0$ for square-integrable functions $\psi(x)$ going to zero at infinity.

9. THE UNCERTAINTY PRINCIPLE

One of the most distinctive properties of the Fourier transformation is that, if the original function is very narrow ($\Delta x \to 0$), its Fourier transform is very wide ($\Delta k \to \infty$). The mathematical idea behind this is quite intuitive: we cannot build a narrow function by only summing over wavelengths ($\lambda = 2\pi/k$) larger than the function's width. In other words, the product $\Delta x \Delta k$ is approximately unity. The precise statement of the mathematical theorem is: $\Delta x \Delta k \geq 1/2$, for all functions ψ and $\partial \psi/\partial x$ differentiable, square-integrable, and going to zero at infinity. A formal proof of this general theorem of the Fourier transformation, independent of the Planck constant, can be attained by following the steps below:

a) Start out from the obvious expression:

$$\int_{-\infty}^{+\infty} \left[\left(x + \lambda \frac{\partial}{\partial x}\right)\psi\right]^* \left(x + \lambda \frac{\partial}{\partial x}\right)\psi \, dx \geq 0 \quad \forall \lambda \in \mathbb{R}$$

where $x + \lambda \frac{\partial}{\partial x} = x + i\lambda k$ is a *real* operator, with $k \equiv -i\frac{\partial}{\partial x}$ (complex).

b) Add and subtract $\psi \cdot (x - \lambda \frac{\partial}{\partial x})$ to the expression in square brackets, and show that

$$\int_{-\infty}^{+\infty} \left[\left(x + \lambda \frac{\partial}{\partial x}\right)\psi^* - \psi^* \cdot \left(x - \lambda \frac{\partial}{\partial x}\right)\right]\left(x + \lambda \frac{\partial}{\partial x}\right)\psi \, dx = 0,$$

because the integrand is actually a full derivative, that verifies

$$\int_{-\infty}^{+\infty} \frac{\partial}{\partial x}\left[\psi^* \cdot \left(x + \lambda \frac{\partial}{\partial x}\right)\psi\right] dx = 0,$$

assuming ψ and $\partial \psi/\partial x$ meet the condition $\left. |x|\psi|^2 + \lambda \psi^* \frac{\partial \psi}{\partial x} \right|_{-\infty}^{+\infty} = 0$.

c) Finally express the remainder of the original integral as

$$\langle x^2 \rangle + \lambda^2 \langle k^2 \rangle - \lambda \geq 0 \quad \forall \lambda \in \mathbb{R}.$$

Assuming $\langle x \rangle = 0$ and $\langle k \rangle = 0$ leads to $\langle x^2 \rangle = (\Delta x)^2$ and $\langle k^2 \rangle = (\Delta k)^2$. The Fourier transform $\mathscr{F}(\partial \psi / \partial x) = ik f(k)$ needs to be used. The theorem proof follows after examination of the discriminant of the above parabola in λ.

We leave as an instructive exercise to the student to work out the details of each of the above steps, and just comment about the transcendental physical significance of this result, when we take into account *De Broglie's relation*: $p = \hbar k$, so that $\Delta p = \hbar \Delta k$ actually means the *dispersion* of the particle's *momentum* distribution.

The result was first stated by the German physicist Werner Heisenberg in 1927, and it is known in physics as position-momentum **uncertainty principle**: if we know with great precision (Δx) the position occupied by a moving body, then large fluctuations are inevitable on the value of its momentum (Δp), their respective wave function widths being related by the inequality

$$\Delta x \Delta p \geq \hbar/2. \qquad (9.1)$$

The above inequality is verified by *every quantum state* represented by the wave function ψ, at all times. Its physics impact is huge. It reveals the impossibility by principle to know simultaneously, *with infinite precision*, the position of a moving body along a given axis, and its momentum on the same axis.

Once $\langle x \rangle = 0$ is set in the derivation above, $k_0 \equiv \langle k \rangle \neq 0$ in general, defined by a factor $e^{-ik_0 x} \psi(x)$. Since $\Delta x \left[(\Delta k)^2 + k_0^2 \right]^{1/2} \geq 1/2 \; \forall k_0 \in \mathbb{R}$, the inequality is also verified for $k_0 = 0$. Therefore $\Delta x \Delta p \geq \hbar/2$ is universally established, being 100% relativistic (as the De Broglie relations are). Of course, it is *also verified for a photon* ($m = 0$) of any frequency ω.

An immediate consequence of expression (9.1) is that every moving body confined to a region of space of size $2\Delta x$ necessarily acquires a (non relativistic) kinetic energy with *mean value*:

$$\langle T \rangle = \frac{1}{2m} (\Delta p)^2 \geq \frac{1}{4} \frac{1}{2m} \frac{\hbar^2}{(\Delta x)^2},$$

if we assume $\langle p \rangle = 0$ and disregard the mean velocity of the particle.

Because what we have calculated is the mean value of T, we are not talking about individual quantum fluctuations of a particular measurement of the kinetic energy, but of a *displacement* of the majority of the measurements.

The $\Delta x \Delta p$ lower bound can be used to estimate the *minimum energy* that the particle acquires, when confined over a distance $2\Delta x$, *no matter the potential $U(x)$ that actually exerts the confining force*. That is, the energy of its ground state (governed by quantum fluctuations, and never far from the $\Delta x \Delta p$ lower bound). Interestingly, in the ultrarelativistic limit the $\Delta x \Delta p$ lower bound goes slightly higher, thus having, for example, for electrons in 1D: $\Delta x \Delta p \geq a\hbar$, with $a \in [\frac{1}{2}, (1+\frac{\sqrt{5}}{2})\frac{1}{3}]$ for $\Delta p/(mc) \in [0, +\infty)$ [1].

Hence for all physical systems, despite some 50% variability in the minimal product $\Delta x \Delta p$, the uncertainty principle may be used in the form of the *approximate equality*: $\Delta x \Delta p \sim \hbar$, to directly write the kinetic energy $T \sim \frac{\hbar^2}{2m(\Delta x)^2}$, or relativistically $T \sim \sqrt{(\hbar/\Delta x)^2 c^2 + m^2 c^4} - mc^2$ [2].

The theorem $\Delta x \Delta p_x \geq \hbar/2$ (1D) can be formulated easily in 3D in the form $\Delta r \Delta p \geq 3\hbar/2$, for in a radial system: $\Delta r = \sqrt{\langle r^2 \rangle} = \sqrt{3} \Delta x$ and $\Delta p = \sqrt{\langle p^2 \rangle} = \sqrt{3} \Delta p_x$. Confinement in 3D creates independent momentum fluctuations in all space coordinates, which add up quadratically.

Thus the uncertainty principle allows us to estimate the energy of a physical system *just from its size*. We suggest here assessing two particularly insightful exercises: one is estimating the proton mass from 3 bound massless quarks $m_q \approx 0$ ($r_p = \sqrt{\langle r^2 \rangle} = 0.84$ fm) and another is deriving the diffraction angle acquired by any particle of momentum p (photon included) that emerges from a hole of diameter d (here the principle should be applied in 1D on the direction perpendicular to p).

When the particle moves in the form of a dispersive packet with *group velocity* v_g, it makes perfect sense to define a `time uncertainty`, arising from its space uncertainty, as $\Delta t \equiv \Delta x / v_g$.

The energy uncertainty can also be derived from Δp as

$$\Delta E = \Delta\left(\frac{p^2}{2m}\right) = \frac{p}{m}\Delta p = v_g \Delta p \, ,$$

where the velocity of the particle is also involved, and we have used the chain rule to relate the variations of energy and momentum.

---

[1] Iwo and Zofia Bialynicki-Birula, NJP 21 073036 (2019), and for photons PRL 108, 140401 (2012).

[2] in many relevant cases, the energy estimate *improves* when a value very close to the lower bound is taken ($\hbar/2$, or $3\hbar/2$ in 3D).

The same result is obtained by differentiating the relativistic expression for the energy $E = \sqrt{p^2c^2 + m^2c^4}$ (with m being the particle's rest mass), taking into account that $dE/dp = v_g = \beta c = pc^2/E$ holds in this case. It is clear that the product $\Delta E \Delta t$ does not depend on v_g anymore, and from expression (9.1) we get the `energy-time uncertainty principle`:

$$\Delta E \Delta t \geq \hbar/2 . \qquad (9.2)$$

To recap, the approximation $\Delta x \Delta p \sim \hbar$ (in 1D) is valid on two conditions:
- the confined particle should be close to its ground state, with energy either $E_0 \sim (\hbar/\Delta x)^2/(2m)$ or $E_0 \sim \sqrt{(\hbar/\Delta x)^2 c^2 + m^2 c^4}$.
- the particle must not have been set in free motion, *after* the confinement. Otherwise, a time-dependent squared width: $\Delta x^2 + \Delta x_t^2$ with extra component $\Delta x_t \approx (\Delta p/m)t$ [3] will set in, while Δp remains constant.

Should the above conditions not be met, the product $\Delta x \Delta p$ will ever increase above $\hbar$, the inequality $\Delta x \Delta p \geq \hbar/2$ ($\Delta r \Delta p \geq 3\hbar/2$ in 3D) being always satisfied.

A numerical estimate of the amount of energy provided by quantum fluctuations can be given for the lightest particle we have in ordinary matter, the electron, with mass $m_e = 9.109 \times 10^{-31}$ kg. Its kinetic energy in eV for the three reference sizes: $\Delta x = 1$ mm, $1\,\mu$m, and 1 Å, yields the respective values: 3.8×10^{-14} eV, 3.8×10^{-8} eV and 3.8 eV. While in the first two cases the energy is unobservable in the laboratory, it becomes quite significant in the third case (the one actually realized in atoms).

When the confining potential is known, the uncertainty principle always allows to estimate (typically in 1D) *both* the energy E_0 of the ground state and the object's size Δx, without having to solve the time independent Schrödinger equation (as will be seen in Chapter 12). This can be done *easily* by finding the *minimum* of the average total energy expressed as $E_0 \sim \langle E \rangle = \langle T \rangle + \langle U \rangle$, as function of Δx.

Expression (9.2) for the energy-time uncertainty can also be used as the approximate equality: $\Delta E \Delta t \sim \hbar$. It allows estimating the kinetic energy ΔE acquired as a consequence of the *time location*, as discussed earlier in Section 2.1. For instance, if $\Delta t \sim h/E$ then this energy becomes significant in relative terms, since $\Delta E/E \sim (\hbar/E)(1/\Delta t) \sim 1/2\pi \sim 16\%$.

---

[3] the exact relativistic expression is $\Delta x_t = (\Delta p/m)t/(1+a)^{3/2}$ with $a = \gamma^2\beta^2 = (p/mc)^2$, as easily shown by differentiating $\Delta x_t = \Delta v t$ from $v = pc^2/\sqrt{p^2c^2 + m^2c^4}$. Note that such *time-dependent* width is always *zero* for a photon propagating in vacuum.

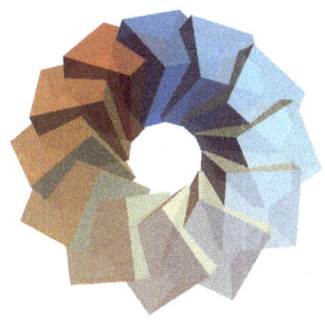

10. EXTENSION TO THREE DIMENSIONS

All of the ideas developed above have been formulated assuming that the motion and the force field $U(x,t)$ occur in one dimension. However, the real motion takes place along the three space coordinates $r = (x,y,z)$. The extension to three dimensions of all of the above can be directly achieved, and we suggest the student to write down correctly, with vector notation **r**(t), the following expressions:

- Feynman's propagation amplitude
- Schrödinger's equation
- wave function
- plane wave, dispersion relation and group velocity
- mean value of a scalar A and of a vector magnitude $\mathbf{A} = (A_x, A_y, A_z)$
- Fourier transformation
- scalar product of wave functions
- p operator
- $H = \dfrac{p^2}{2m} + U$ operator
- uncertainty principle

The normalization factors in the Fourier transformation and in the Feynman propagator need to be changed from $(2\pi)^{1/2}$ to $(2\pi)^{3/2}$. We shall write below the result for Schrödinger's equation in 3D:

$$i\hbar\frac{\partial \psi}{\partial t} = \left(\frac{-\hbar^2}{2m}\left(\frac{\partial^2}{\partial x^2} + \frac{\partial^2}{\partial y^2} + \frac{\partial^2}{\partial z^2}\right) + U(\mathbf{r},t)\right)\psi. \qquad (10.1)$$

Chapter 10. EXTENSION TO THREE DIMENSIONS

Note that the Euclidean geometry, and the Pythagorean theorem, become involved in Feynman's propagator, and this is why the kinetic energy operator can immediately be extended to 3D.

Using the Laplace operator $\Delta \equiv \frac{\partial^2}{\partial x^2} + \frac{\partial^2}{\partial y^2} + \frac{\partial^2}{\partial z^2}$, the equation is commonly writen as:

$$i\hbar \frac{\partial \psi}{\partial t} = \left(\frac{-\hbar^2}{2m} \Delta + U(\boldsymbol{r},t) \right) \psi . \tag{10.2}$$

Or even simpler, using the Hamiltonian operator $H = \frac{p^2}{2m} + U(\boldsymbol{r},t)$:

$$i\hbar \frac{\partial \psi}{\partial t} = H\psi . \tag{10.3}$$

We have already seen the 3D formulation of the uncertainty principle. It should be kept in mind that position and momentum are intrinsically correlated only along the *same coordinate* $\Delta x_i \Delta p_i \sim \hbar$ with $x_i = x, y, z$. Recall that confinement in 3D generates independent momenta in each dimension, that are added quadratically.

The 3D Feynman propagator reads as follows:

$$\psi(\boldsymbol{r},t) = \iiint K(\boldsymbol{r},t;\boldsymbol{r}_1,t_1) \, \psi(\boldsymbol{r}_1,t_1) d^3 \boldsymbol{r}_1 . \tag{10.4}$$

The inverse Fourier transform in 3D is written as:

$$\psi(\boldsymbol{r}) = \frac{1}{(2\pi)^{3/2}} \iiint f(\boldsymbol{k}) \, e^{i\boldsymbol{k}\cdot\boldsymbol{r}} d^3 \boldsymbol{k} . \tag{10.5}$$

The most relevant case that makes the 3D analysis unavoidable is that of the angular momentum $\boldsymbol{L} = \boldsymbol{r} \times \boldsymbol{p}$, that we shall undertake in Chapter 15. Mean values and dispersion of the $\boldsymbol{L} = (L_x, L_y, L_z)$ operators can be calculated as 3D integrals, although knowledge of their eigenstates will generally simplify the calculations, as will be seen in Chapter 16.

11. EIGENSTATES AND MEASURABLE VALUES

Given a measurable magnitude A, we ask ourselves the question: are there wave functions ψ such that A is well defined on them, i.e. a repetition of the measurements on the state ψ will always throw the same real value $a \in \mathbb{R}$? In a more precise way: are there wave functions ψ, such that $\Delta A = 0$ on them? We are now ready to answer this question mathematically, according to the definition of ΔA we gave in (8.2). Let us assume we are able to solve the *eigenvalue problem* for the operator A: $A\psi = a\psi$. This means finding the possible eigenvalues a and eigenvectors ψ for the above equation [1]. Then expression (8.3) allows us to verify immediately that indeed $(\Delta A)^2 = a^2 \langle \psi | \psi \rangle - a^2 \langle \psi | \psi \rangle^2 = 0$ on the eigenvectors, since $\langle \psi | \psi \rangle = 1$ according to the normalization property.

So the problem has been stated in mathematical terms: we have to solve the eigenvalue problem for the operator A. The states ψ looked for are then the corresponding eigenvectors, called in Quantum Mechanics eigenstates.

We know the eigenvalues a need to be real, although in general they do not fill up the real line, but define a subset $\mathbb{A} \subset \mathbb{R}$ therein. In many cases this subset, which is called *spectrum* of the operator A, will be discrete (isolated points). All real values that do not belong to the spectrum of A ($a \notin \mathbb{A}$) will be forbidden, and will never be measured in the laboratory.

Furthermore, the measurements performed on the same eigenstate will always throw the same value, hence being *reproducible* [2].

---

[1] which will be, in general, a partial differential equation.

[2] this makes possible that experiments of extremely high precision can be achieved in physics. Often such experiments are a driving factor in physics development, or constitute a major technological breakthrough. In any case, measurements of great precision can be attained on the *mean values* $\langle A \rangle$.

Chapter 11. EIGENSTATES AND MEASURABLE VALUES

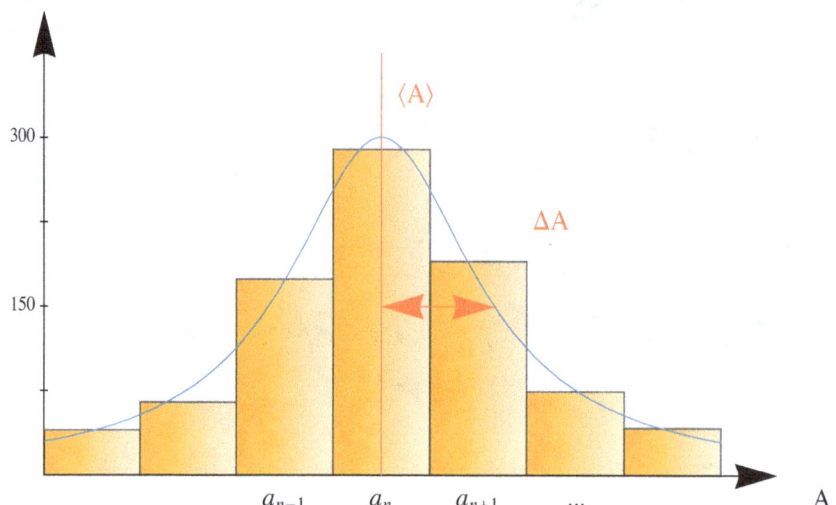

Figure 11.1: *Distribution of 1100 laboratory measurements of the magnitude A. The initial state ψ is not an eigenstate of the operator A, and has been prepared to be the same everytime. The measurements are randomly distributed, according to a calculable probability density (as $|\langle \psi_n | \psi \rangle|^2$). Note that only the eigenvalues a_n of A are measurable.*

Let it be noted that in order to determine the spectrum, it is essential to impose the condition that the wave function must be continuous in $\mathbb{R}^3$, and square-integrable.

Just as eigenvectors of a *Hermitian* matrix corresponding to different eigenvalues are orthogonal to each other, the same is true with the eigenstates of an operator: they form an orthonormal set of functions (the so-called eigenfunctions), according to our definition of the scalar product.

Furthermore, the eigenstates corresponding to the same eigenvalue a_n form a subspace of dimension $N(a_n)$ within the Hilbert space, in which it is also possible to define an orthonormal basis $\{\psi_{n,m}, m = 1, N(a_n)\}$, where $N(a_n)$ is the multiplicity of the eigenvalue a_n.

Let us assume we perform a measurement of the magnitude A on an initial state ψ which is *not* an eigenstate, and obtain the value $a \in \mathbb{R}$. It is generally admitted that the wave function has changed after the measurement, and must now be an eigenstate ψ_a, corresponding to the measured value.

Since the measurement excludes alternative values $a_x \neq a$ [3] which have not been realized, these should no longer be considered as part of the particle's state of motion. This phenomenon by which the realization of a measurement randomly alters the state of the moving body is known in the literature as the `collapse of the wave function`.

This idea was originally put forward by the Hungarian mathematician John Von Neumann in 1932, and it is today generally accepted in Quantum Mechanics. The collapse of the wave function takes place in the laboratory in a sudden and unpredictable way, just as we described earlier in Section 2.6 for the detection of a photon. It is just the way the non deterministic character of Quantum Mechanics is manifest, such that full knowledge of the wave function does not allow us to know a priori which will be the measured eigenvalue. An example of how such measurements can be distributed is illustrated in Figure 11.1.

It should be noted that the time evolution of the wave function $\psi \to \psi_a$ during the measurement `is not governed by Schrödinger's equation`. As of today, it is *unknown* what are the physical laws behind the collapse of the wave function. The above is easily understood, since the resulting wave function using the Schrödinger equation would be uniquely determined from ψ by the Feynman propagator, according to (10.4).

It can be shown that (as in the analogous case with matrices), the eigenfunctions of any Hermitian operator A form an orthonormal basis in the Hilbert space. Hence every wave function, at a given time, can be expanded as

$$\psi(r) = \sum_n \sum_m c_{n,m} \psi_{a_n,m}(r) , \qquad (11.1)$$

where $\{\psi_{a_n,m}(r), m = 1, N(a_n)\}$ is the orthonormal set of eigenfunctions corresponding to each different real eigenvalue a_n, where the complex coefficients $c_{n,m}$ are obtained as `scalar products` $c_{n,m} = \langle \psi_{a_n,m} | \psi \rangle$. The normalization property demands that $\sum_{n,m} |c_{n,m}|^2 = 1$, which implies that the quantities $P_n = \sum_m |c_{n,m}|^2$ actually determine the *probability that the measurement made on ψ throws the value a_n*.

The collapse of the wave function means that all coefficients $c_{n,m}$ that are *incompatible* with the measurement are instantaneously removed, or *erased*, from $\psi(r)$ ($c_{n,m} = 0$) after the measurement.

---

[3] we are not considering here the statistical fluctuations inherent to the experimental measurement, thus implicitly assume no statistical or systematic uncertaintly.

Chapter 11. EIGENSTATES AND MEASURABLE VALUES

A well known example of the above ideas is found in the modulus $|L|$ of the angular momentum $L = r \times p$, where the index n takes the values $l = 0, 1, \cdots \infty$, and its Z-component L_z illustrates the multiplicity index m, taking values $m_l \in (-l, \cdots +l)$, as we shall see in Chapter 16, after solving the eigenvalue problem.

Note that the *inverse Fourier transform* we saw in Chapter 10 can be regarded as yet another example of the previous idea of expansion over eigenstates. Indeed let us write

$$\psi(r) = \frac{1}{(2\pi)^{3/2}} \iiint f(k) \, e^{ikr} d^3k = \sum_k c_k \, e^{ikr} = \sum_n c_{k_n} \, e^{ik_n r} .$$

The above equivalent expressions can be seen as a particular case of (11.1). First, the 3D Riemann integral is actually the limiting case of a *sum* over a discrete (integer) index n, the multiplicity index m being absent. The sum runs over a set of *eigenvalues* $a_n = p_n = \hbar k_n$ and corresponding *eigenfunctions* (De Broglie plane waves $e^{ik_n r}$) of the *momentum* operator p. Second, the momentum components p_n may indeed nearly fill up the whole real momentum space $\mathbb{R}^3$, or may just represent a small set of momentum values. In the latter case, $f(k)$ needs to be represented by a Dirac delta function. Whatever the case, the complex coefficients project over the eigenstates: $c_{k_n} = f(k_n) = \langle \psi_{a_n} | \psi \rangle = \langle e^{ik_n r} | \psi \rangle / (2\pi)^{3/2} \equiv \langle k_n | \psi \rangle$ and do no more than performing a sampling of the Fourier transform of $\psi(r)$.

The Dirac notation introduced above for the momentum eigenstates $|k\rangle \equiv e^{ikr}/(2\pi)^{3/2}$, motivates an interesting expression for the Fourier transformation as a scalar product (or projection):

$$f(k) = c_k = \langle k | \psi \rangle = \frac{1}{(2\pi)^{3/2}} \langle e^{ikr} | \psi \rangle$$

where $e^{ikr*} = e^{-ikr}$ is to be taken into account.

12. THE STATIONARY STATES

Let us imagine a periodic classical trajectory of energy E [1], starting from r_0 with period Δt, such as the one represented in Figure 12.1. The particle goes over the points of this trajectory as a function of time, and we can associate to each point r the value of the phase given by the Feynman propagator $K = Ae^{iS/\hbar} = Ae^{i(S_0 - Et)/\hbar} = e^{-iEt/\hbar}Ae^{iS_0/\hbar}$, where S is the classical action $S = \int_{t_0}^{t} L dt$ [2]. If we want the motion to retain the periodic character it has in Classical Mechanics, this phase must return, at every point r on the trajectory, to its original value after a closed orbit, which implies the condition

$$\frac{S_0}{\hbar} = 2\pi n \quad n = 0, 1, 2, \ldots \infty.$$

In order to apply the above condition to integrable systems, it is important to know that trajectories are confined to *invariant tori* in phase-space [3], whose projection onto the coordinate space has a *boundary*.

---

[1] in Classical Mechanics with N freedoms, periodic motion is well characterized for both integrable and chaotic systems. In the former case, it should be called more properly *multiperiodic*, since the motion is confined to *invariant tori* in the phase-space of constant energy E, with $N-1$ independent frequencies. Since rational numbers are a dense set in real numbers, we can always make these frequencies proportional to appropriate integers, so that they are commensurable, and the motion is truly periodic. In chaotic systems no such tori exist, because there are no constants of motion, other than the total energy. Yet periodic orbits are also meaningful, even if they are unstable after many cycles.

[2] propagator that becomes exact at all times, when used over the classical trajectory.

[3] it was Einstein in 1917, in one of his least known articles, who first noted the existence of such tori, A. Einstein Verb Dtsch Phys Ges 19: 82 (1917), as refered by M. Gutzwiller in his book "Chaos in Classical and Quantum Mechanics", p. 208, Springer 1990.

Chapter 12. THE STATIONARY STATES

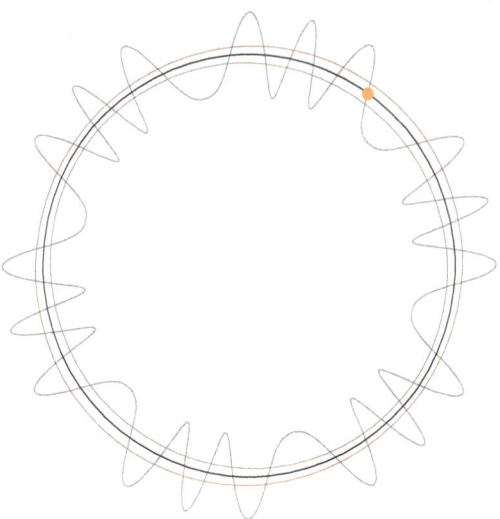

Figure 12.1: *Illustration of the idea of a stationary state, using closed classical trajectories (black and brown): quantum fluctuations are of such nature that at each point on the trajectory (orange) the phase of the propagator (wavy line) returns to exactly the same value after one cycle. Many classical orbits of this kind need to be considered to form a stationary state.*

We need to make sure that the action is a single-valued function of the coordinates, which requires that a phase loss of $\pi/2$ has to be admitted [4] everytime the projection of the closed trajectory reaches the above boundary of the coordinate space [5].

Thus the previous quantization condition needs to be corrected to yield: $S_0/\hbar = n2\pi + \mathcal{M}\pi/2 = (n+\mathcal{M}/4)2\pi$, where $\mathcal{M}$ is the (positive integer) number of boundary crossings, called earlier the Maslov index, reading $S_0 = (n+\mathcal{M}/4)h$, just the quantization condition formulated in Chapter 2.

For chaotic systems, periodic trajectories are also enhanced when they are in phase, but no mutually commensurable periods exist, and many trajectories, with very long periods, may contribute to the same energy [6].

---

[4] to achieve a thorough comprehension of this statement in Classical Mechanics, some dedicated reading is advised. See, for instance, M. Gutzwiller, "Chaos in Classical and Quantum Mechanics", Springer 1990, pp. 657-665, and also S. C. Creagh, J. M. Robbins and R. G. Littlejohn, Physical Review A Vol. 42, No. 4 (1990).

[5] a relevant example of the above is the motion of a particle in an attractive central potential of the form $U(r) = \beta r^d$ ($d > -2$), where the radial coordinate r runs back and forth between the limits (r_{min}, r_{max}). The boundary is reached precisely at these limits, when the sign of the radial momentum is reversed.

[6] a comprehensive development of this idea will be found in M. Gutzwiller, "Chaos in Classical and Quantum Mechanics", pp. 301-305, Springer 1990.

Whether the motion is integrable or chaotic, we may conjecture that states of well defined energy do exist in Quantum Mechanics, where the time factor $e^{-iEt/\hbar}$, introduced by the propagation over the classical trajectory, is inherited by the wave function. This would be natural, since this factor indicates a *frequency* in accord with De Broglie's relation $\omega = E/\hbar$, thus uniquely determined by the energy.

Hence we see that, under the assumption that the classical trajectories govern the exact path integral calculation, the postulate of quantized reduced action comes out naturally from Feynman's propagation, and formula $S_0 = \oint p d r = (n+\alpha)h$ indeed emerges as an excellent approximation to calculate the allowed energies for integrable systems. In fact, we have already used it quite effectively in two particular cases (quantum oscillator and Hydrogen atom). It is known generically in the literature as the semiclassical approximation.

In any case, having an approximate solution is not enough, and we want to have the *exact* solution to the energy problem, since Feynman's theory is exact. But even more importantly, most physical systems of interest (such as multielectronic atoms, molecules, atomic nuclei, or condensed matter in solids or liquids), are chaotic in their classical formulation, which prevents the semiclassical approximation above from being useful on them. Hence the most effective way to proceed is to use Schrödinger's equation, and realize that, in the periodic motion, what we want is the particle's wave function $\psi(r,t)$ to have a time-dependent common periodic phase at all space points. Which amounts to saying that $\psi(r)$ should remain invariant under the action of the Feynman propagator.

The above phenomenon is analogous to that of normal modes for waves. Thus we want the probability density at each point r to be constant in time

$$\frac{\partial |\psi(r,t)|^2}{\partial t} = 0 \quad \forall r \in \mathbb{R}^3 , \tag{12.1}$$

that is, a stationary probability density, like an unchanging temperature distribution in a thermal medium, or a fluid velocity distribution in a stationary regime.

Chapter 12. THE STATIONARY STATES

We can attain solutions of the above type to the Schrödinger equation $i\hbar\frac{\partial \psi}{\partial t} = H\psi$ by pursuing the two steps below:

1) Find solutions to the eigenvalue problem of the Hamiltonian operator

$$\left(\frac{-\hbar^2}{2m}\Delta + U(r)\right)\psi(r) = E\psi(r) \qquad (12.2)$$

for some real value of the parameter E (total energy), and impose the condition that $\psi(r)$ makes sense as a wave function. Specifically, it must accomplish:
   a) The normalization condition $\int |\psi(r)|^2 d^3r = 1$, that requires a convergent integral ($< +\infty$).
   b) The continuity of the wave function at each space point $r \in \mathbb{R}^3$. Also the continuity of the directional derivatives ($\frac{\partial \psi}{\partial r}$), unless we have assumed an infinite potential at that particular point [7].

It is actually the verification of the two conditions above what really causes a restriction on the energy values E, bringing them to their quantized values.

2) Build the following time dependent wave function

$$\psi(r,t) = e^{-iEt/\hbar}\psi(r), \qquad (12.3)$$

which indeed represents the desired stationary states, for
   a) it is a solution of the Schrödinger equation $i\hbar\frac{\partial \psi}{\partial t} = H\psi$, and
   b) the condition (12.1) is met.

We leave as a simple exercise to check that the wave function (12.3) verifies both the above statements.

Equation (12.2) can be written as

$$H\psi = E\psi, \qquad (12.4)$$

which is called `time independent Schrödinger's equation` in the literature. The operator H is specified by the force field $U(r)$, and the exact calculation of the energies of the stationary states is thus reduced to a mathematical problem of eigenvalues, which implies a second order partial differential equation, independent of time. The eigenvalues must be real, of course, since H is self-adjoint.

---

[7] such potentials (impenetrable walls, Coulomb law at the origin, Dirac delta functions, etc.) are never fully realizable in practice, but their introduction may greatly simplify the problem in question, and are customarily used.

Incident or outgoing plane waves (or spherical waves), in a given spatial direction, can be enforced as boundary conditions. More genuine boundary conditions may also be imposed, where the wave function is required to be zero at some surface (impenetrable walls), which is defined by setting there an infinite potential.

Of course, all the results obtained from equation (12.4), with appropriate boundary conditions, bear all the precision of the Feynman path integral we saw in 3.1. As in every eigenvalue problem, according to what was stated in Chapter 11, the eingenstates meet the condition $\Delta E = 0$, that is, their energy is well defined. Solving the differential equation $H\psi = E\psi$ for the most elementary force fields in one and several dimensions (central fields, Coulomb potential, particle in a box, piece-wise potentials, barriers, harmonic oscillator, etc) will be dealt with in chapters 17, 20, 23, 24, 25, 26, and 27.

Once the energy eigenvalue problem has been solved for a given potential field, the set $\{\Phi_{n,m}(r), E_n, n = 1, \cdots, \infty, m = 1, \cdots N(E_n)\}$ formed by its eigenfunctions and associated eigenvalues is available to us, where m runs over the quantum numbers necessary to describe the different orthogonal eigenfunctions of equal energy E_n, in the subspace associated with that particular eigenvalue [8].

The *dimension* of the above subspace (maximum number of orthogonal states of equal energy), is called quantum degeneracy: $g_n \equiv N(E_n)$ in the literature. Even if the phenomenon of degeneracy is also there in Classical Mechanics (being, in general, infinite), the key issue is that, in the physical reality, it is always a calculable *positive integer* number.

The complete set of states $\{\Phi_{n,m}(r), E_n, g_n, n = 1, \cdots \infty, m = 1, \cdots g_n\}$ defines the exact solution for the energy levels, and level density, that every physical system has, when periodically confined.

The *entropy* of a statistical system formed by many particles is just the natural logarithm of the total number of orthogonal states $\Delta \mathcal{N} = g(E)\Delta E$ that the system may have, within the interval ΔE in which its internal energy fluctuates: $S = k_B \ln(g(E)\Delta E)$. The Bolzmann constant k_B provides its proper JK^{-1} dimension.

---

[8] as an example, for the particle of kinetic energy E_n enclosed in a cubic box, that we shall analyse in Chapter 23, the possible orientations of its momentum p. For a particle subject to a central potential, those necessary to define its angular momentum state: (l,m) that shall be described in Chapter 16.

Expansion in stationary states

Eigenvectors of unit norm, with different eigenvalues, of a *Hermitian* matrix, always form an orthonormal set, within the vector space where this matrix operates. Therefore every wave function at a given time can be expressed as a linear combination of energy eigenstates. If, for the sake of simplicity, we leave out the degeneracy index m, which is not essential for the discussion below, we can write: $\psi(r) = \sum_n c_n \Phi_n(r)$, with *complex* numbers $c_n \in \mathbb{C}$ such that $\sum_n |c_n|^2 = 1$.

Taking into account Schrödinger's equation from Eq. (12.3) at each of the above terms, the time dependence of the wave function is given by the expansion

$$\psi(r,t) = \sum_n c_n e^{-iE_n t/\hbar} \Phi_n(r) . \tag{12.5}$$

It is important to realize that such function is not, in general, a stationary state, unless all coefficients c_n are zero, except one. In other words, the necessary and sufficient condition for a state to have a well defined energy is that "it does not move", according to the definition given by expression (12.1). Thus we can say that the *motion* originates in physics from the *superposition of different energies* (frequencies), a conception that is alien to Classical Mechanics.

Since the scalar product $c_n = \langle \Phi_n | \psi \rangle$ is independent of t, we can write

$$\psi(r_2,t_2) = \sum_n \langle \Phi_n | \psi \rangle e^{-i\frac{E_n}{\hbar}(t_2-t_1)} \Phi_n(r_2) = \sum_n \left(\int \Phi_n^*(r_1) \psi(r_1) d^3 r_1 \right) \Phi_n(r_2) e^{-i\frac{E_n}{\hbar}(t_2-t_1)} .$$

By reordering the above factors, and recalling how the propagator operates on the wave functions, as $\psi(r,t) = \int K(r,t;r_1,t_1)\psi(r_1,t_1)d^3r_1$, we conclude that the *exact* Feynman propagator can be written as

$$K(r_2,t_2;r_1,t_1) = \sum_n \Phi_n(r_2) \Phi_n^*(r_1) e^{-i\frac{E_n}{\hbar}(t_2-t_1)} , \tag{12.6}$$

where the sum runs over the set of *all* stationary states. The propagation can be considered forward in time ($t_2 > t_1$) or backward in time ($t_2 < t_1$), when using the above expression. Note that in order to perform the full reverse propagation in time $(r_2,t_2) \to (r_1,t_1)$, the *conjugate* propagator $K^*(r_2,t_2;r_1,t_1)$ is required.

The above formula is of great utility in all kinds of problems, including particle scattering under the action of a given potential, or time-dependent perturbation theory. It also shows that the stationary states of well defined energy are spatially invariant under the action of the propagator, as we conjectured at the beginning of this chapter. To see this, just let K operate on an eigenstate $\Phi_m(r)$ according to (10.4), and take into account their orthonormality.

The time evolution operator

Expresssion (12.5) provides us with an alternative procedure to determine the exact time evolution of the wave function $\psi(r,t)$ from its initial value $\psi(r,0)$, that complements the one we already had (10.4) with Feynman's propagator.

The new procedure does not require a convolution, instead it involves the action of a *linear operator* $\mathscr{U}(t)$ in the Hilbert space, acting on the initial wave function as $\psi(r,t) = \mathscr{U}(t)\psi(r,0)$. Assuming we have previously solved the energy eigenvalue problem, and that we are in possession of the orthonormal base $\{\Phi_n(r), E_n, n = 1, \infty\}$, where for simplicity we have omitted the degeneracy index, it is obvious that the operator $\mathscr{U}(t)$ takes, in that particular base, the form of the following diagonal matrix:

$$\mathscr{U}(t) = \mathrm{diag}\left(e^{-\frac{i}{\hbar}E_1 t}, e^{-\frac{i}{\hbar}E_2 t}, \cdots\right).$$

Assuming the Hamiltonian H is constant in time, it also takes the form of a (constant) diagonal matrix in that base:

$$H = \mathrm{diag}\left(E_1, E_2, \cdots\right).$$

Thus the linear operator $\mathscr{U}(t)$ can be expressed in *exponential form*: $\mathscr{U}(t) = e^{-\frac{i}{\hbar}Ht}$, taking into account that the exponential of a diagonal matrix is itself diagonal. Which, using the standard Dirac notation $|\cdots\rangle$ for the quantum states, is formulated as:

$$|\psi(r,t)\rangle = \mathscr{U}(t)|\psi(r,0)\rangle = e^{-\frac{i}{\hbar}Ht}|\psi(r,0)\rangle. \qquad (12.7)$$

The above expression remains valid in any other base, in which the Hamiltonian adopts the non-diagonal form $H' = U^{-1}HU$, owing to the property $e^{-i(U^{-1}HU)t/\hbar} = U^{-1}e^{-iHt/\hbar}U$, that can easily be derived from analysis of the series expansion of the exponential.

Equation (12.7) plays a central role in Quantum Mechanics. The operator $\mathscr{U}(t)$ is called in the literature time evolution operator, and it is easily checked that it is a *unitary* operator satisfying: $\mathscr{U}^+ = \mathscr{U}^{-1}$, where $\mathscr{U}^+$ denotes the conjugate and transpose matrix ($\mathscr{U}^+\mathscr{U} = \mathbb{1}$). The operator is unitary because it conserves the *norm* of the wave function at all times t: $\int |\psi(r,t)|^2 d^3r = \langle \psi|\psi \rangle = 1$, in perfect concordance with its probabilistic interpretation.

When the number of energy levels (dimension of the Hilbert space of states) is *finite*, the Hamiltonian is just a Hermitian matrix H of dimension N, and the problem of finding the stationary states is no other than that of *diagonalising* it, that is, determining the aforementioned unitary matrix U, constituted by its eigenvectors.

13. THE BOHR FORMULA

As seen, time-dependence is generated by a linear combination of stationary states, according to Schrödinger's equation, such that the wave function is no longer stationary, but takes the form of a moving pulse.

Let us analyze the probability density when the particle finds itself in a state which is a superposition of two different energy levels: E_2 (*high* state) and E_1 (*low* state), with $E_2 > E_1$. Such case is quite general and happens in a great deal of physical systems, causing a phenomenon called quantum oscillation. We shall see that the particle travels virtually across the space in a periodic way, with a frequency ω determined by Bohr's formula: $E_2 - E_1 = \hbar\omega$. In case the particle has an electric charge, the oscillation gives rise to the emission of *one* photon of the same frequency after a certain time Δt, in qualitative agreement with the prediction of Classical Electrodynamics. Yet Bohr's formula is not just a mere consequence of energy conservation and the existence of photons (given that *two* photons could also be emitted with frequencies $\omega_1 \neq \omega_2$), but it is actually rooted in Schrödinger's equation, as will be seen below.

Indeed, let us assume $\psi = c_1\psi_1 + c_2\psi_2$, with $\psi_{2,1}$ being the wave functions of the *high* and *low* states, respectively, with $c_{1,2} \in \mathbb{C}$. Since the overall phase does not have any physical meaning, we may assume without loss of generality that c_1 is real positive ($c_1 > 0$), and that $c_2 = |c_2|e^{i\phi_0}$, ϕ_0 being their relative phase, with $|c_1|^2 + |c_2|^2 = 1$.

If we single out a given space point r, the probability density at that point gets a contribution from the interference between both states:

$$|\psi(r,t)|^2 = |c_1\psi_1 + c_2\psi_2|^2 = |c_1|^2|\psi_1|^2 + |c_2|^2|\psi_2|^2 + 2\operatorname{Re}\left(|c_1c_2||\psi_1\psi_2|e^{-i\left(\frac{(E_2-E_1)t}{\hbar} - \Phi\right)}\right),$$

where $\Phi(r) \equiv \phi_2(r) - \phi_1(r) - \phi_0$ depends on the phase difference at that point between the wave functions $\psi_1 \equiv |\psi_1|e^{i\phi_1(r)}$ and $\psi_2 \equiv |\psi_2|e^{i\phi_2(r)}$. It can easily be checked that the probability density is a periodic function of time:

$$|\psi(r,t)|^2 = A(r) + B(r)\cos(\omega t - \Phi(r)). \tag{13.1}$$

The particle density then oscillates between a maximum value $A+B$ (constructive interference) and a minimum value of $A-B$ (destructive interference) with a frequency ω that is independent of the point r, and determined by Bohr's formula. An example is shown in Figure 13.1, that we shall discuss below.

The functions A and B do depend on r, and take the positive values $A(r) = |c_1|^2|\psi_1(r)|^2 + |c_2|^2|\psi_2(r)|^2$ and $B(r) = 2|c_1c_2||\psi_1(r)\psi_2(r)|$. It is then clear that the probability increase at that point necessarily means a decrease at other points, due to the wave function normalization, and conversely. Therefore what happens is a global displacement of the particle (oscillation) of periodic character.

The oscillation stops instantly once the particle is localized, or its energy is measured. Note that, during the oscillation process, the energy is not well defined ($\Delta E \neq 0$). It is a useful exercise to calculate ΔE explicitely (dispersion of the H operator), and show that $\Delta E = |c_1c_2|(E_2 - E_1)$.

The energy measurement, for charged particles, is achieved by detecting the emitted photon, which provides evidence that the particle is already in the *down* state. In this case, the time Δt elapsed before photon emission can be estimated on average from the uncertainty principle $\Delta E \Delta t \sim \hbar$ [1].

A good 3D example of oscillations, of the charged type, is found when the electron in a Hydrogen atom adopts a quantum state which is a superposition of the ground state with $n=1$ (1s orbital) and the first excited state with $n=2$, $l=1$ and $m_l=1$ (2p orbital). These wave functions are obtained by solving the time-independent Schrödinger equation with the Coulomb field, and will be studied in Chapter 20. In spherical coordinates, they have the generic form: $\psi = R_{nl}(r)Y_l^{m_l}(\theta,\phi)$.

---

[1] the time Δt should not be confused with the oscillation period $T \equiv 2\pi/\omega$.

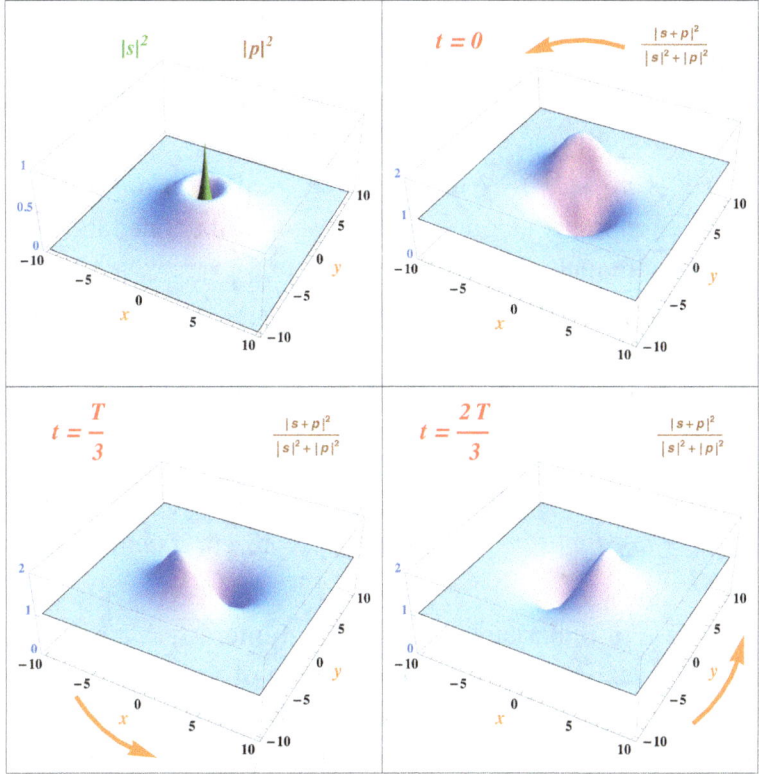

Figure 13.1: *Time dependent probability density in the horizontal plane (XY) of the non stationary state that results from superposition of the orbitals $2p$ $(n = 2, l = 1, m_l = 1)$ and $1s$ $(n = 1, l = 0)$ of Hydrogen, divided by the sum of probabilities. It can be seen how the interference between them generates an electric dipole which is a function of time, that rotates around the proton with Bohr's frequency (period $T = \frac{2\pi}{\omega}$). The horizontal scales show units of the Bohr radius a_0. In the top left panel the $2p$ orbital has been multiplied by a factor of 50.*

The electron density in the horizontal plane is depicted in Figure 13.1 as a function of time, where it can be seen how it rotates around the proton, as a consequence of the interference. For the sake of clarity, $c_1 = c_2 = \frac{1}{\sqrt{2}}$ has been assumed, and the density has been divided by the sum of electron densities in both orbitals, also represented in Figure 13.1.

This phenomenon is called spontaneous radiation and the quantum superposition of the wave functions actually occurs due to the action of the electromagnetic field of a vacuum at that particular frequency (recall Section 2.6, formula (2.7)). The average lifetime is calculable in *Quantum Electrodynamics* from the wave functions. Some of its basic principles, referring to Larmor's radiation law, shall be given in Chapter 27.

There are also examples of quantum oscillation with particles that are electrically neutral. Most strikingly, the relativistic case of `neutrino flavor oscillations` that happen with neutrinos of a given momentum p and two different rest masses $m_1 \neq m_2$ (hence having different energies $E_{1,2} = \sqrt{p^2c^2 + m_{1,2}^2 c^4}$). What stops the oscillation in this case is the neutrino detection in the laboratory, showing a well defined flavor through its interaction with matter [2].

All of the above cases of oscillation are generically expressed in the form

$$|\psi(t)\rangle = e^{-i\omega_1 t}\left(|\psi_1\rangle + e^{-i(\omega_2-\omega_1)t}|\psi_2\rangle\right),$$

where the first factor can never contribute to the spatial density $|\psi(r)|^2$, the oscillation being manifest in the second term, with its characteristic frequency $\omega = \omega_2 - \omega_1$ (the symbols $|\psi_{1,2}\rangle$ represent stationary states, with $\omega_{1,2} = E_{1,2}/\hbar$).

Another case of remarkable technological interest, that belongs to the same category, deals with the synchronization of `atomic clocks` of very high precision. An external electromagnetic actuator causes the quantum superposition above to take place in the moving atoms of a gas system, thus initiating their internal oscillation. The process is stopped after a certain time by a second actuator, synchronized with the previous one by means of an external oscillator [3]. As in the preceding cases, the stopping of the oscillation is caused by the detection process, now embodied by an atom detector only sensitive to the *up* state. Extremely small phase shifts (10^{-14} or even better), between the external oscillator and the atomic clock, can be measured in this way. GPS satellites contain multiple atomic clocks that contribute very precise time data to the GPS signals.

---

[2] to handle this particular case, you may assess the exercise proposed in the course.
[3] see a more complete description in the book "The Quantum World", M. Le Bellac, World Scientific (2014), along with the exercise proposed in the course, dedicated to this topic.

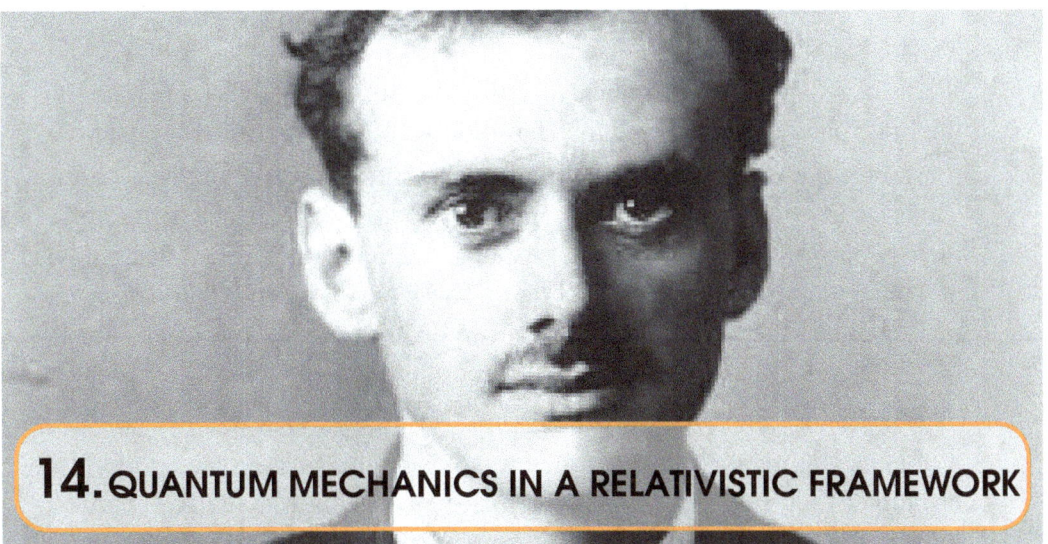

14. QUANTUM MECHANICS IN A RELATIVISTIC FRAMEWORK

In a number of cases, we have addressed the study of quantum physics under the assumption that the particle velocity is small compared with the light velocity c, although all previous ideas were historically born from a relativistic approach.

The reason not to deal with the full relativistic formalism in this core course is twofold:

- On a general basis, Schrödinger's equation $i\hbar \frac{\partial \psi}{\partial t} = H\psi$ is itself fully relativistic, provided a proper definition of the H operator is adopted. For the electron, this is embodied in the *Dirac equation*. Free-motion solutions are still plane waves (called *spinors*), which have an extra degree of freedom (the electron *spin*), with the additional potentiality of representing the antiparticles (the *positron*), closely related with their spatial propagation properties backward in time. The formalism requires somewhat lengthier calculations.

- Most applications in atomic physics, molecular physics, condensed matter physics, and even nuclear physics, involve non relativistic velocities, and the calculations with the Dirac equation are unnecessary (except for precision phenomena, such as vacuum polarization). In addition, the handling of many-body interactions becomes much more complicated. Only in particle physics is the relativistic formalism widespread.

Chapter 14. QUANTUM MECHANICS IN A RELATIVISTIC FRAMEWORK

Let us recapitulate and realize that, in all the formalism thus far seen, only the Feynman propagator and the Hamiltonian of the Schrödinger equation (based on the equation $E = \frac{p^2}{2m}$) were expressed in non relativistic form. This was in contrast to the De Broglie relations, the uncertainty principle, the orthonormal basis of stationary states, the Schrödinger equation itself, and the unitary time evolution operator. These are all essentially relativistic concepts, and their related formalism can be used in all kinds of problems involving velocities comparable to c, after making correct use of relativistic kinematics.

Let us simply add that a detailed relativistic analysis of the uncertainty principle, focused on the product $\Delta p \Delta t$, reveals that it actually becomes more stringent [1]: the complete localization of a particle in space turns out to be impossible, no matter its momentum uncertainty. The maximum spatial location (minimal Δx) that a particle of rest mass m can have, is given in relativity by its so-called `Compton wavelength` [2]:

$$\lambda_C = \frac{h}{mc},$$

and no experiment can determine the position of a particle with greater precision than the above. For the electron, $\lambda_C = 0.02426\,\text{Å} = 2426\,fm$. This is conceivably the most far-reaching implication of relativity in Quantum Mechanics.

It should be pointed out that, in a fully relativistic framework, Quantum Mechanics is required to describe particle-antiparticle pair creation and annihilation, and this approach means being immersed in a more general formulation, called Quantum Field Theory. This theory extends Quantum Mechanics also in applications not necessarily relativistic, in condensed matter physics. In electromagnetism, it is called Quantum Electrodynamics.

---

[1] see the preeminent discussion in "Relativistic Quantum Theory I" by V.B. Berestetskii, E.M. Lifshitz and L.P. PitaevskiiI, Landau Course of Theoretical Physics, Pergamon Press (1971), pp. 1-4.

[2] a simple way to see this is associating the best attainable time location $\Delta t \sim h/(mc^2)$, that we saw in Section 2.1, with $\Delta E \sim \hbar/\Delta t = mc^2/(2\pi)$, that amounts to $\Delta p = mc/(2\pi)$. Thus a best spatial resolution may be inferred: $\Delta x \sim \hbar/\Delta p = h/(mc)$.

15.   THE ANGULAR MOMENTUM

Every moving body that occupies the space point r with momentum p *rotates* about the origin of the coordinate system, as long as both vectors are not collinear. Rotation is called in physics *angular momentum* and it is described by the cross product $J = r \times p$. In the International System of units (SI) it is measured in $J \cdot s$. Its vector character expresses the fact that every instantaneous rotation of the space can be decomposed into three rotations about each of the coordinate axes: $J = (J_x, J_y, J_z)$. When the nature of the rotating particle is *known* by its mass m (even if it is zero, as for a photon), it is called *orbital* angular momentum, and denoted by L. Rotation of a particle upon itself is measured in the same units, and it is called in physics *spin* or *intrinsic* angular momentum, being denoted by S. In its most general form, the angular momentum of a moving body can be written as $J = L + S$.

Note that the angular momentum is measured in the same units as the Planck constant (units of action), hence it is the most specifically quantum observable that we may find in physics. Given its definition in terms of the position and momentum vectors, it is clear that the uncertainty principle will have a great deal to say about the intrinsic limitations of performing precision measurements of its components.

Since the wave function $\psi(r)$ encodes *all* the information about a body's state of motion, it must also provide detailed account of its orbital angular momentum at any time t. The internal degrees of freedom that allow to describe its spin, which are *necessarily* of quantum nature, as we shall see, are translated into the presence of extra components of its wave function $\psi_i(r)$.

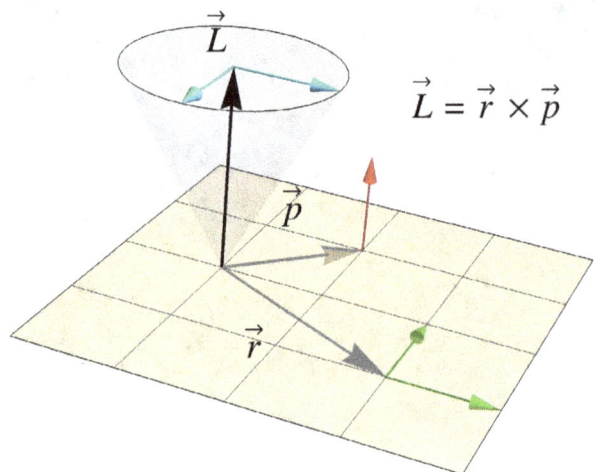

Figure 15.1: *Implication of the uncertainty principle on the angular momentum. Simultaneous and perpendicular quantum fluctuations of the momentum and position vectors preclude a precise definition of **L** in 3D.*

The exact implication of the uncertainty principle in the measurement of angular momentum is quite simple: *it becomes impossible to perform independent and infinitely precise measurements of its three components* (J_x, J_y, J_z). This conclusion can be reached easily by *reductio ad absurdum*, as shown in Figure 15.1. Though we have chosen **L**, the reasoning is equally valid for the spin. Let us assume we have full knowledge of the three components of angular momentum, expressed by the vector $\boldsymbol{L} \neq 0$. Then we can align the Z-axis with this vector, and be sure that both the **r** and **p** vectors lay in the plane perpendicular to the Z-axis. But this means we know at the same time, with infinite precision, the position occupied by the particle along the Z-axis ($z = 0$) and its momentum along the same direction ($p_z = 0$), in open violation of the uncertainty principle, that demands $\Delta z \Delta p_z \geq \hbar/2$.

As a consequence of the above, angular momentum can never be fully projected onto any space direction, as quantum fluctuations over the Z-axis of the **p** vector are necessarily correlated with quantum fluctuations of the **r** vector in the perpendicular plane (through their cross product), as illustrated in Figure 15.1. Thus the **L** vector undergoes inevitable fluctuations over its two perpendicular directions.

The aforementioned fluctuations impart the body an *additional* kinetic energy of rotation, such that the following *strict inequality* is fulfilled (for $L \neq 0$):

$$|L| > L_z . \qquad (15.1)$$

For no wave function, at any given time, may independent measurements of L^2 and L_z^2 throw equal values. The above inequality signals once again a profound difference between Classical Mechanics (where the equality $|L| = L_z$ *is* possible) and the physical reality. It has a deep implication in atoms, molecules, atomic nuclei, and elementary particles, as we shall see.

In particular, the above inequality is key to understand the phenomenon of paramagnetism, that relies on the fact that the individual magnetic moments of the atoms (which are a consequence of their internal angular momentum) can never be fully aligned in the direction of an external magnetic field. The limitation does not only arise from thermal fluctuations, but fundamentally from quantum fluctuations.

Let us recall that in Classical Mechanics force fields with a central potential $U(|r|)$ originate trajectories that are always confined in a plane. This is a consequence of the fact that the 3 components of the L vector, which stands perpendicular to the trajectory, are constant in time. The presence of quantum fluctuations does not violate the conservation of L on *average*, but obviously prevents the motion to be confined in a plane.

To make the above ideas more precise, it becomes unavoidable to find the eigenstates of angular momentum, and associated eigenvalues. Given that $p = -i\hbar \nabla$, there is no ambiguity as to what mathematical form the angular momentum operators must take in Quantum Mechanics. They should read as follows:

$$L = -i\hbar \begin{vmatrix} i & j & k \\ x & y & z \\ \frac{\partial}{\partial x} & \frac{\partial}{\partial y} & \frac{\partial}{\partial z} \end{vmatrix} = -i\hbar \left[\left(y\frac{\partial}{\partial z} - z\frac{\partial}{\partial y} \right) i + \left(z\frac{\partial}{\partial x} - x\frac{\partial}{\partial z} \right) j + \left(x\frac{\partial}{\partial y} - y\frac{\partial}{\partial x} \right) k \right].$$

Just because angular momentum plays a specific role in central potentials, it is convenient to use spherical coordinates [1], to express the wave function $\psi(r) = \psi(r, \theta, \phi)$ and the (L_x, L_y, L_z) operators.

---

[1] the spherical coordinates establish a 1−1 relationship $(x, y, z) \leftrightarrow (r, \theta, \phi)$ within $\mathbb{R}^3$ through the standard transformation: $x = r\sin\theta\cos\phi$, $y = r\sin\theta\sin\phi$, and $z = r\cos\theta$, with the volume element transformation $d^3r = r^2 d(\cos\theta) d\phi dr$.

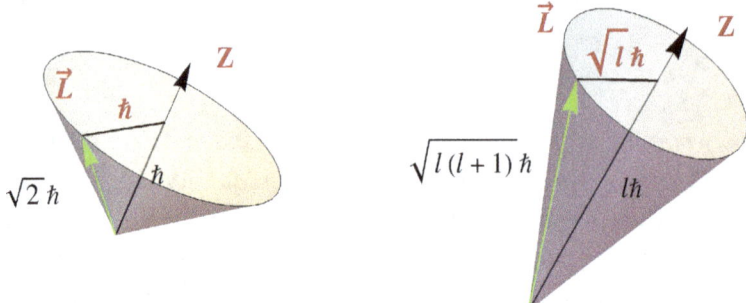

Figure 15.2: *Maximal projection ($m_l = +l$) of the **L** vector in the eigenstates of angular momentum, for $l = 1$ (left), and $l = 10$ (right). Quantum fluctuations preclude its full orientation over any axis, being maximal for $l = 1$. For larger values of l ($l \to \infty$), the **L** vector is more and more focalized, with ever smaller opening angle over the Z-axis.*

The **p** vector may be orthogonally decomposed in the plane perpendicular to **L**, see Figure 15.1, as the sum of two vectors: one proportional to **r** and another perpendicular to it, $\boldsymbol{p} = p_r \boldsymbol{u}_r + p_t \boldsymbol{u}_t$ (with $\boldsymbol{u}_L = \boldsymbol{u}_r \times \boldsymbol{u}_t$ being the unit vector along **L**, and $|\boldsymbol{u}_r| = |\boldsymbol{u}_t| = 1$). Upon $\boldsymbol{L} = \boldsymbol{r} \times \boldsymbol{p}$, only the second component contributes, thus we have $|\boldsymbol{L}| = r p_t$.

Accordingly, the kinetic energy of a moving body of mass m (here in the non relativistic limit) can always be decomposed into two orthogonal components:

$$T = \frac{p^2}{2m} = \frac{p_r^2}{2m} + \frac{p_t^2}{2m} = \frac{p_r^2}{2m} + \frac{L^2}{2mr^2}, \qquad (15.2)$$

which are called *radial* kinetic energy (the particle moves towards or away from the origin) and *rotational* kinetic energy, respectively. Note that the balance between them is not universal, but depends on where the *origin of coordinates* is located, in order to define the rotation (naturally at the center of force, when there is one).

The fact that the rotational kinetic energy $T_R = L^2/(2mr^2)$ becomes infinite at very short distances, for a given angular momentum, is well known in mechanics. This energy is due to the centrifugal force. It is clear that measuring T_R is equivalent to a measurement of the *square* of the modulus of the angular momentum vector $|\boldsymbol{L}|$, when the radius of gyration is fixed.

Additionally, we can perform measurements of the body's rotation about each of the coordinate axes (L_x, L_y, L_z), by means of some kind of mechanical device, even for individual atoms or particles. Assuming the Pythagorean theorem ensures the relationship $L^2 = L_x^2 + L_y^2 + L_z^2$.

Let us now write these operators in spherical coordinates, including L^2, by applying the chain rule for partial derivatives. The simplest case is:

$$L_z = -i\hbar \frac{\partial}{\partial \phi} \; .$$

The complete analogy of this operator with the momentum operator along a given axis: $p_z = -i\hbar \frac{\partial}{\partial z}$ is truly suggesting, when we realize that the azimuthal angle ϕ actually represents the rotation angle. Note that angles are dimensionless, thus reminding us that the dimension of the operator is precisely that of an action ($J \cdot s$).

The angular momentum operators in the perpendicular directions come out to be:

$$L_x = -i\hbar \left(-\sin\phi \frac{\partial}{\partial \theta} - \frac{\cos\phi}{\tan\theta} \frac{\partial}{\partial \phi} \right) \quad L_y = -i\hbar \left(\cos\phi \frac{\partial}{\partial \theta} - \frac{\sin\phi}{\tan\theta} \frac{\partial}{\partial \phi} \right).$$

The operator $L^2 = L_x^2 + L_y^2 + L_z^2$ is quadratic, thus involving second derivatives, with the result:

$$L^2 = -\hbar^2 \left[\frac{1}{\sin\theta} \frac{\partial}{\partial \theta} \left(\sin\theta \frac{\partial}{\partial \theta} \right) + \frac{1}{\sin^2\theta} \frac{\partial^2}{\partial \phi^2} \right]. \qquad (15.3)$$

We can express the total kinetic energy by making use of the well known form of the Laplace operator Δ in spherical coordinates, as:

$$T = \frac{p^2}{2m} = \frac{-\hbar^2}{2m} \Delta = \frac{-\hbar^2}{2m} \left(\frac{\partial^2}{\partial r^2} + \frac{2}{r} \frac{\partial}{\partial r} \right) - \frac{\hbar^2}{2mr^2} \left[\frac{1}{\sin\theta} \frac{\partial}{\partial \theta} \left(\sin\theta \frac{\partial}{\partial \theta} \right) + \frac{1}{\sin^2\theta} \frac{\partial^2}{\partial \phi^2} \right].$$

It is interesting to realise that both expressions above are related as:

$$T = H_0 = \frac{-\hbar^2}{2m} \left(\frac{\partial^2}{\partial r^2} + \frac{2}{r} \frac{\partial}{\partial r} \right) + \frac{L^2}{2mr^2}, \qquad (15.4)$$

since the operator L^2 is precisely the angular part of the Laplace operator. This is actually a physical consequence of the earlier seen decomposition of the kinetic energy (15.2). Its verification in the language of operators comes to confirm the consistency of their quantum definition (and incidentally, it allows us to read off the correct definition of p_r^2, in brackets).

For a given initial wave function $\psi(r)$, it is possible to make individual measurements of each of the 3 components of angular momentum, and build a vector L in $\mathbb{R}^3$ space with them. According to the previous discussion about the uncertainty principle, these measurements will be distributed over a three dimensional point cloud which may never be narrowed down to a single point.

We can predict the center of that cloud (its mean value) by calculating, in spherical coordinates, each of the following 3D integrals:

$$\langle L \rangle = \langle \psi | L | \psi \rangle = \begin{pmatrix} \int \psi^*(r)(L_x\psi)d^3r \\ \int \psi^*(r)(L_y\psi)d^3r \\ \int \psi^*(r)(L_z\psi)d^3r \end{pmatrix}.$$

The above *real* vector is uniquely determined by the wave function, and represents the *mean value of the angular momentum* of that body. We may also use the integral calculus to determine the dispersion of the previous measurements around the mean values on each axis: $\Delta L_z^2 = \langle L_z^2 \rangle - \langle L_z \rangle^2$, etc. On the other hand, it is possible to perform a direct measurement of L^2 through the kinetic energy of rotation. Hence we have four quantities that can be measured independently, namely: L_x, L_y, L_z, and L^2.

It is thus imperative to solve the eigenvalue problem for these operators, to determine the allowed quantum values of the measurements, as well as the eigenfunctions associated to them in each case. Which means solving four partial differential equations, of the type: $L_i\psi = l_i\psi$ ($i = x, y, x$), and $L^2\psi = a^2\psi$, where l_i y a^2 are real numbers.

In order to avoid unnecessary calculations, let us first answer some key questions: will it be possible to find a wave function $\psi(r)$ simultaneously fulfilling the first 3 equations above, for $i = x, y, z$? Obviously not, since that would imply full knowledge of the 3 components of L, that goes against the uncertainty principle. Furthermore: may the set of eigenvalues found for L_z be any different from that found for L_x and L_y? Clearly not, either, since the choice of the Z-axis in the spherical coordinates is arbitrary, the 3 space coordinates being indistinguishable.

It is also easy to convince oneself that simultaneously solving *any two* of the above differential equations, with two perpendicular axes, is equally unworkable. Just take the system formed by $L_z\psi = l_z\psi$ and $L_x\psi = l_x\psi$ and realize that the tentative factorized function $\psi = A(\theta)e^{il_z\phi/\hbar}$ cannot be a solution, since $\partial A/\partial \phi = 0$ must be met.

Therefore, it only makes sense trying to solve the following *two* coupled differential equations:

$$L_z \psi(r) = l_z \psi(r) \qquad L^2 \psi(r) = a^2 \psi(r) , \qquad (15.5)$$

that together determine the modulus of L and its projection L_z onto a given axis, as shown in Figure 15.2.

Before proceeding to the mathematical analysis of the above system of equations in the next chapter, we may test our physical intuition and conjecture what would be the possible solutions to the eigenvalue problem, i.e. the quantum values of $|L|$ and L_z. Given that angular momentum is measured in the same units as h, from $L_z = -i\hbar \frac{\partial}{\partial \phi}$, we may predict the allowed quantum values to be $L_z = \pm\hbar, \pm 2\hbar, \cdots$, where the sign $\pm$ indicates dextro/levo rotation. Still the question comes: is *zero* an allowed quantum value? And even more importantly: may the allowed quantum values for $|L|$ be *the same* as the absolute value of those found for L_z?

It is clear that the inequality $|L| > L_z$ will make the above impossible. What would then be the *minimum non zero* value of $|L|$, and subsequent quantum values? Should the action quantization in the axes perpendicular to Z also operate as multiples of $\hbar$, that would give us a minimum value of $\sqrt{2}\hbar$, or perhaps $\sqrt{3}\hbar$. The detailed solving of the above equations will certainly provide us with the correct answer. We anticipate here the result:

$$|L| = 0, \sqrt{2}\hbar, \sqrt{6}\hbar, \sqrt{12}\hbar, \sqrt{20}\hbar, \cdots \qquad (15.6)$$

16. THE ANGULAR MOMENTUM EIGENSTATES

To deal with the eigenvalue problem of the system of equations (15.5), we shall use the method of performing a power series expansion of the solutions, and finding a recursive law for its coefficients. The following Lemma will allow us to identify the divergent solutions, by comparing them with known functions. It is derived easily from repeated application of the Stolz-Cesaro theorem for the convergence of successions, which we shall not take the time to prove here.

LEMMA OF ASYMPTOTIC SIMILARITY: *The functions $f(x) = \sum_n a_n x^n$ and $g(x) = \sum_n b_n x^n$ are defined within the radius of convergence $x \in (-c, c) \subset \mathbb{R}$. If $\lim_{x \to \pm c} g(x) = (\pm)\infty$ and $\lim_{n \to \infty}[(a_{n+1}/a_n)/(b_{n+1}/b_n)] = 1$, then the function $f(x)$ is also divergent, and it meets the asymptotic behaviour defined by $\lim_{x \to \pm c} f(x)/g(x) = C$, with $C \in \mathbb{R}, C \neq 0$.*

Let us study the solutions to the first of equations (15.5), and search for factorized wave functions $\psi(r\theta\phi) = R(r)A(\theta)B(\phi)$, with $l_z \in \mathbb{R}$. Given the form of the operator $L_z = -i\hbar \frac{\partial}{\partial \phi}$, it is obvious that only the function $B(\phi)$ will be constrained, and we may take for $R(r)$ and $A(\theta)$ arbitrary functions, provided the normalization property is fulfilled $\langle \psi | \psi \rangle = 1$. In particular, the condition $\int_0^\infty r^2 |R(r)|^2 dr = 1$ should be satisfied.

The solution for $-i\hbar \partial B/\partial \phi = l_z B$ is well known:

$$B(\phi) = e^{\frac{i}{\hbar} l_z \phi} ,$$

despite the fact that the above function is not continuous in $\mathbb{R}^3$, but shows instead a cut (or step) along the positive X-axis.

Indeed this is because the function limits are different in general for $\phi \to 0$ (right), and $\phi \to 2\pi$ (left), unless the condition $B(0) = B(2\pi) = 1$ is satisfied. Which automatically constrains the measurable values of l_z:

$$l_z = m\hbar \ , \ m = 0, \pm 1, \pm 2, \cdots$$

We have thus shown that the previous conjecture is true, of measurable values of L_z being signed multiples of $\hbar$ (reduced action quantization), having also elucidated that the solution $m = 0$ is indeed *allowed* in Quantum Mechanics. It is customary to use the letter m (m_l when necessary) to denote the above integers. Of course, the minus sign corresponds to counter-clockwise rotations about the Z-axis.

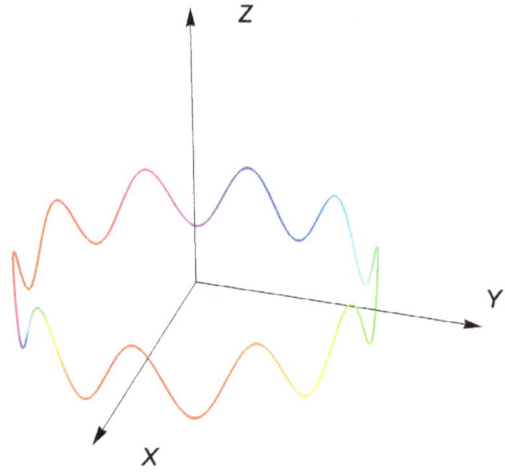

Figure 16.1: *Variation of the phase of the wave function when the particle moves in a normal circle centered on the rotation axis (Z), at any radial distance.*

Let us observe the structure of the factorized eigenfunctions in spherical coordinates:

$$\psi = R(r)A(\theta)\,e^{im\phi}.$$

They show an oscillating *phase* when we sample the function over normal circles centered on the rotation axis, the oscillation being faster with higher m values, see Figure 16.1. Every space dependent phase in a wave function means an *energy*, according to the Feynman propagator. In this case it is the *kinetic energy of rotation*, since the variation of the potential would be zero in a central field. Note, however, that this phase has no effect on the $|\psi|^2$ value, hence the probability density of finding the particle in normal circles centered on the Z-axis remains rigorously constant.

Keeping the above solutions for a given value of m, i. e. of L_z, we may now find the proper dependence on θ so that the complete system of equations (15.5) is fulfilled. This will allow us to mathematically determine the sequence of values for the modulus $|L|$ that are compatible with L_z.

Using expression (15.3) for the L^2 operator, the eigenvalue problem can be stated as: $L^2(A(\theta)B(\phi)) = a^2 A \cdot B$, with $a^2 \in \mathbb{R}$. After dividing by $B(\phi)$ on both sides:

$$\frac{1}{\sin\theta}\frac{\partial}{\partial\theta}\left(\sin\theta\frac{\partial A}{\partial\theta}\right) + \frac{A}{\sin^2\theta}\left(\frac{1}{B}\frac{\partial^2 B}{\partial\phi^2}\right) + \frac{a^2}{\hbar^2}A = 0.$$

We now substitute the solutions $B = e^{im\phi}$, taking into account the second derivative $\partial^2 B/\partial\phi^2 = -|m|^2 B$. Notice that all information about the sign of m has then been lost, after differentiation. For this reason we must write hereafter $|m|$ instead of m, and understand that every conclusion drawn for $|L|$, as well as for the $A(\theta)$ function itself, will be independent of the direction of rotation (clock- or anti-clock wise), as it was to be expected.

We obtain an ordinary second order differential equation for the $A(\theta)$ function:

$$\frac{1}{\sin\theta}\frac{\partial}{\partial\theta}\left(\sin\theta\frac{\partial A}{\partial\theta}\right) - |m|^2\frac{A}{\sin^2\theta} + \frac{a^2}{\hbar^2}A = 0, \qquad (16.1)$$

that it is conveniently expressed as a function of the argument $x \equiv \cos\theta$, by replacing $A(\theta)$ with $y(x)$, from $\partial/\partial\theta = -\sin\theta\,\partial/\partial x$ (of course, these x and y standard variables have nothing to do with the cartesian coordinates with the same name):

$$(1-x^2)y'' - 2xy' + \left(\frac{a^2}{\hbar^2} - \frac{|m|^2}{1-x^2}\right)y = 0,$$

which is known in textbooks as the *associated* Legendre equation, well studied since the 19th century.

It is important to realise that *we are not interested in all solutions* to this equation, but only in those defined in the interval $x \equiv \cos\theta \in [-1,+1]$, hooked up with the spherical coordinates. Furthermore, they should give rise to a function ψ acceptable in Quantum Mechanics, i.e. non-singular and square-integrable in $\mathbb{R}^3$. Which means $A(\theta) = y(x)$ cannot be infinite on the Z-axis (the rotation axis), defined by $x \to \pm 1$ (North and South poles). However the structure of the equation is such that we are to expect divergent solutions precisely at these two points.

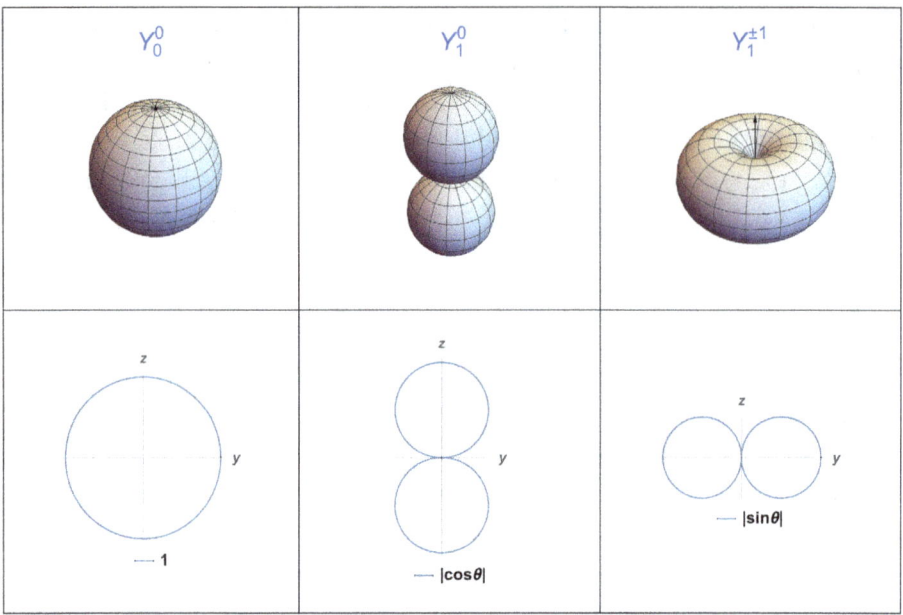

Figure 16.2: *3D representation of the lowest spherical harmonics l = 0 and l = 1, using Wolfram Mathematica 11. The value of $|Y_l^m|$ is proportional to the distance to the origin. The projection of the particle's angular momentum over the vertical axis is $m\hbar$. The lower panels show their projection onto the YZ plane (polar diagrams), where the functions that have been represented are also indicated.*

Aiming to characterize this kind of divergent solutions, we now make a *change of function*: $y(x) \equiv (1-x^2)^{|m|/2} v(x) = \sin^{|m|}\theta\, v(x)$. Which does not imply any restriction of generality, since once the equation is solved for $v(x)$, we recover the wave function $y(x)$ from the above relation. On application of the chain rule, it is straightforward to find the new equation for $v(x)$:

$$(1-x^2)v'' - 2(|m|+1)xv' + \left(\frac{a^2}{\hbar^2} - |m|(|m|+1)\right)v = 0. \qquad (16.2)$$

We now use the method of establishing a recursive law on the coefficients a_n of the power expansion around $x = 0$: $v(x) = \sum_{n=0}^{\infty} a_n x^n$. The recursive law is established upon differentiation of the above series (up to a maximum of two times), grouping together the terms with equal powers of x, and equating all of their coefficients to zero.

The result of these laborious, but elementary, operations, leads to the expression:

$$\frac{a_{n+2}}{a_n} = \frac{(n+|m|)(n+|m|+1) - \frac{a^2}{\hbar^2}}{(n+1)(n+2)}, \qquad (16.3)$$

where each coefficient is obtained as a function of the preceeding one by *two units*. This means the power series is determined by the first two coefficients: a_0 y a_1, which play the role of the two constants of integration that every second order differential equation must have.

Note the expansion contains two separate interleaved series, sharing the same recurrence law: the *even* powers, proportional to a_0, and the *odd* powers, proportional to a_1. Each of them generate a family of solutions, that can be activated independently, by setting one of these parameters to zero.

Technically, the differential equation is solved, since we may code the recursive law in a computer and plot out the solutions for $y(x)$. But we want to know a priori when the solutions are divergent for $x \to \pm 1$, as we cannot plot them all. The Lemma stated at the beginning of this chapter then comes into play. We need to find a known test function that is divergent for $x \to +1$, whose power expansion around $x=0$ has a recursive law equal to the above for $n \to \infty$. As stated by the Lemma, it is the high powers of the expansion that govern the asymptotic behaviour of the function in the limit $x \to +1$ [1].

The test function is taken to be the following, for $m > 0$ [2]:

$$(1-x)^{-m} = 1 + mx + \frac{m(m+1)}{2!}x^2 + \frac{m(m+1)(m+2)}{3!}x^3 + \cdots$$

that is a well known series (Newton's binomial formula), whose recursive law can easily be checked to be:

$$\frac{a_{n+2}}{a_n} = \frac{(n+m)(n+m+1)}{(n+1)(n+2)}.$$

---

[1] it can easily be shown, using the limit test for series, that the solution diverges for $x \to +1$, for values $m > 1/4$. However, we are interested in characterizing the type of divergence that arises in the wave function $y(x)$, and for that purpose using the test function procedure.

[2] for $m = 0$ the given expression is no longer a series, but it can be replaced in all the subsequent reasoning by the function: $-\ln(1-x) = 1 + x^2/2 + x^3/3 + \cdots$, also divergent for $x \to +1$, and with recursive law: $a_{n+2}/a_n = n/(n+2)$.

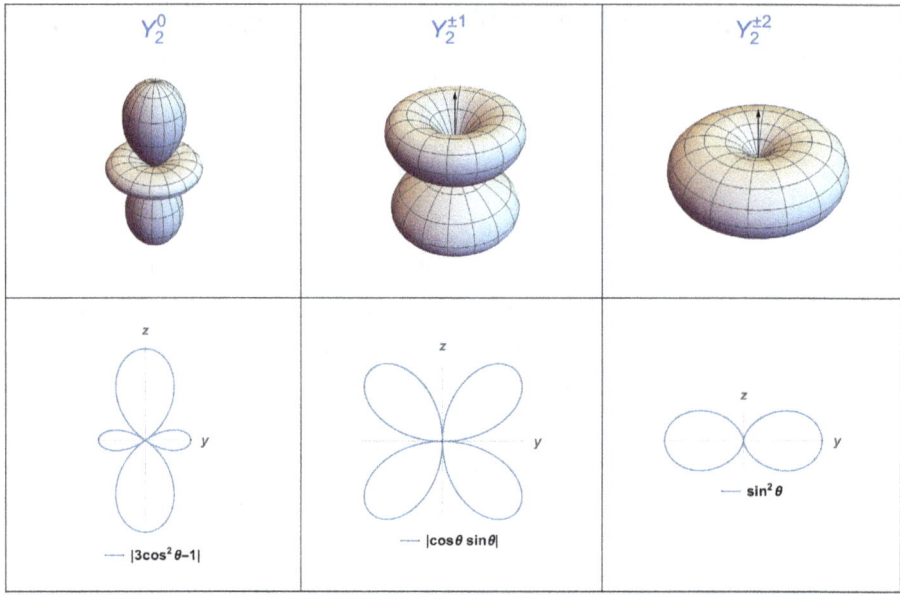

Figure 16.3: *3D representation of the full set of spherical harmonics with $l = 2$, using Wolfram Mathematica 11. The value of $|Y_l^m|$ is proportional to the distance to the origin. The projection of the particle's angular momentum over the vertical axis is $m\hbar$. The lower panels show their projection onto the YZ plane (polar diagrams), where the functions that have been represented are also indicated.*

Let us compare the above expression with formula (16.3), and identify the integer $m > 0$ of one with the $|m|$ of the other, to see that, in the limit $n \to +\infty$, both expressions are identical. Thus the conditions stated in the Lemma are met, and the series (16.3) is unequivocally divergent for $x \to +1$ as $(1-x)^{-|m|}$.

Hence we are led to the surprising conclusion that all solutions to the differential equation (16.2) are divergent for $x \to +1$. Let us recall however the function change we made $y(x) \leftrightarrow v(x)$, and that it is actually the wave function $y(x)$ that needs to be finite. Using the symbol $\sim$ to denote equal asymptotic behaviour for $x \to +1$, we have:

$$y(x) \sim (1-x)^{-|m|}(1-x^2)^{|m|/2} = (1+x)^{|m|/2}(1-x)^{-|m|/2} ,$$

thus the factor $(1-x^2)^{|m|/2} \to 0$ is not able to cure the divergence, and the wave function itself $\psi(r,\theta,\phi)$ becomes discontinuous in $\mathbb{R}^3$, $y(x)$ becoming infinite on the rotation axis.

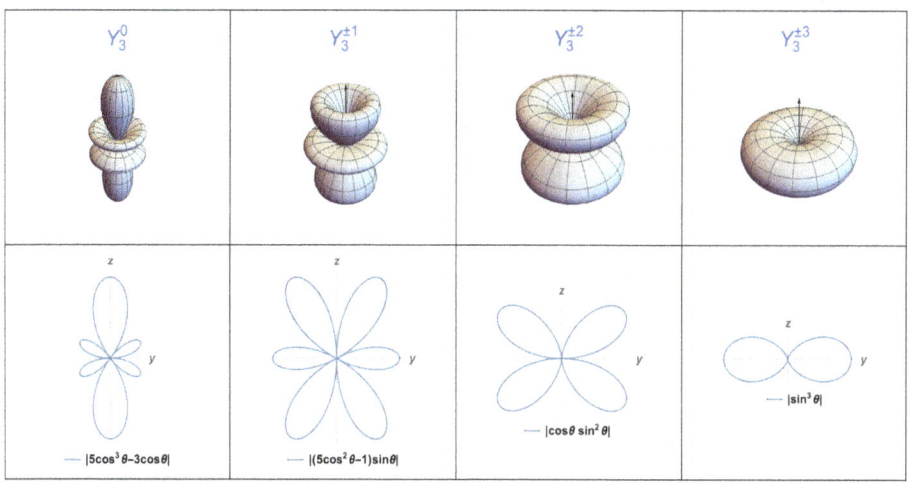

Figure 16.4: *3D representation of the full set of spherical harmonics with $l = 3$, using Wolfram Mathematica 11. The value of $|Y_l^m|$ is proportional to the distance to the origin. The projection of the particle's angular momentum over the vertical axis is $m\hbar$. The lower panels show their projection onto the YZ plane (polar diagrams), where the functions that have been represented are also indicated.*

The postulates of Quantum Mechanics for ψ are violated for a good reason: it makes no sense that electrons rotating about the Z-axis fixate their probability density on the rotation axis, irrespective of the angular momentum value L_z.

The solution to this apparent failure of the idea of eigenstates in Quantum Mechanics comes from detailed observation of the recursive law (16.3). What happens if the value of $a^2/\hbar^2$ is not just any real number, but the product of two consecutive integers? In such case, the series degenerates into a polynomial. The auxiliary Lemma is then not applicable, since it requires the summation of the series to proceed to infinity. Given that the general solution contains two interleaved series, with even and odd powers, we make the key observation that their degeneration into a polynomial can only happen in *one of them*. The other must be cancelled by setting to zero one of the integration constants (either $a_0 = 0$ or $a_1 = 0$). The exact condition is that, for some integer $p = 0, 1, 2 \cdots \infty$, we have:

$$\frac{a^2}{\hbar^2} = (p + |m|)(p + |m| + 1) \, ,$$

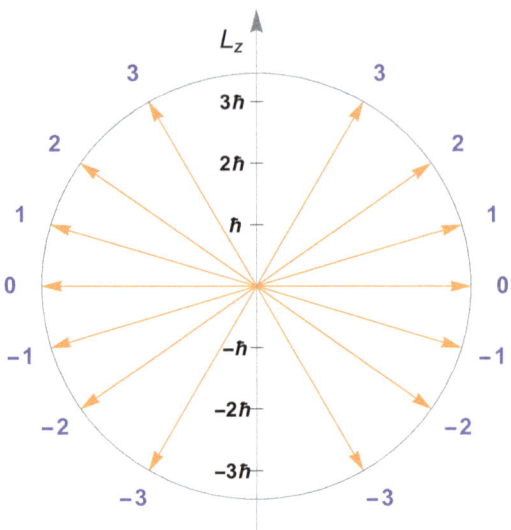

Figure 16.5: *Representation of the quantum values $m = -l, \cdots, +l$ for $l = 3$. This diagram ignores, for the sake of simplicity, the effect of the uncertainty principle, but it precisely illustrates the values of L_z. It constitues an alternative representation to that of Figure 16.4. In the Periodic Table, it characterizes the chemical elements from Cesium to Lutetium, and from Thorium to Lawrencium (f-orbitals).*

Equivalently, it states that the square of the angular momentum $a^2 = L^2$ takes the following values, in units of $\hbar^2$:

$$L^2 = l(l+1)\hbar^2 \qquad l = |m|, |m|+1, \cdots \infty \; , \qquad (16.4)$$

where $l = p + |m|$. Another most popular way to express the above is that, for a fixed value of the positive integer l, we get the following $2l + 1$ allowed values for m, with $L_z = m\hbar$:

$$m = -l, -l+1, \cdots, 0, \cdots +l \; . \qquad (16.5)$$

We have thus obtained *two hugely relevant physical laws*, expressed by equations (16.4) and (16.5). They impact all over the fields of physics and chemistry, from the structure of the Periodic Table of the elements, which is a direct consequence of the latter, to the fundamentals of magnetism, and related technologies.

The sequence of values obtained for $|\boldsymbol{L}|$ confirms what was anticipated at the end of the previous chapter: $|\boldsymbol{L}| = 0, \sqrt{2}\hbar, \sqrt{6}\hbar, \sqrt{12}\hbar \cdots$. When $L_z \neq 0$, the above sequence cannot begin by zero, and with a lowest value $|L_z| = \hbar$, $|\boldsymbol{L}| = \sqrt{1(1+1)}\hbar$ has appeared as the minimum modulus.

In summary, we have seen that concurrent eigenfunctions exist for the L^2 and L_z operators, and that these eigenfunctions are acceptable as quantum states, i.e. they are true eigenstates. Both the modulus of the angular momentum $|\boldsymbol{L}|$, and its projection onto a given axis L_z, are well defined on them. We have also determined the complete set of possible eigenvalues for both magnitudes, which turn out to be in correspondance with specific pairs of integer numbers (l,m).

We saw in Figure 15.2 a representation of the above states, showing the spatial orientation of the $\boldsymbol{L}$ vector: quantum fluctuations take place on a circle, at the edge of a cone. For every pair of integers (l,m) the opening angle of the cone is uniquely determined, and takes quantum values itself: $\theta = \text{ArcCos}(m/\sqrt{l(l+1)})$, with $m = -l, \cdots +l$ (see them represented in Figure 16.5). The maximum orientation of the $\boldsymbol{L}$ vector (minimum θ) compatible with quantum fluctuations, is attained for $m = +l$, the choice depicted in Figure 15.2, for $l = 1$ and for $l = 10$. From a value as high as $\theta = 45°$ for $l = 1$, the opening angles progressively diminish towards zero in the *classical limit*, characterized by $l \to \infty$ and $m \to +\infty$.

The classical limit is thus reached when the angular momentum is high compared to $\hbar$: $|\boldsymbol{L}| \gg \hbar$ and $m \gg 1$, and in that case we get the sensation that the gyration is smooth, with no pitching. Which is never actually true, as clearly manifest in the atomic scale.

The polynomials that meet the recursive law (16.3) for $m = 0$ (with the quantum values $a^2/\hbar^2 = l(l+1)$), are no other than the Legendre polynomials $P_l(x)$, widely documented in the literature. The eigenfunctions we have obtained are given the name of *spherical harmonics*, being denoted by the standard symbol $Y_l^m(\theta\phi)$. Their interest goes far beyond Quantum Mechanics, since they show up, as we have seen, in every solution with radial symmetry to the Poisson equation.

The spherical harmonics are *complex*, non divergent, and infinitely differentiable functions, defined on the unit sphere, that constitute the angular part of the wave function of a rotating body with given values of L^2 and L_z: $\psi = R(r)Y_l^m(\theta\phi)$.

Table 16.1: Legendre polynomials up to $l = 5$

$P_0 = 1$	$P_1 = x$
$P_2 = \frac{1}{2}(3x^2 - 1)$	$P_3 = \frac{1}{2}(5x^3 - 3x)$
$P_4 = \frac{1}{8}(35x^4 - 30x^2 + 3)$	$P_5 = \frac{1}{8}(63x^5 - 70x^3 + 15x)$

The complete mathematical structure of the spherical harmonics, after undoing the function change $y(x) \leftrightarrow v(x)$, and putting back the information of the azimuth ϕ, is the following:

$$Y_l^m(\theta\phi) = s\, N_{lm}\, \sin^{|m|}\theta\, \frac{d^{|m|}}{dx^{|m|}} P_l(x)\, e^{im\phi} \quad (m \geq 0), \quad (16.6)$$

where the Legendre polynomials must be differentiated $|m|$ times (recall $x \equiv \cos\theta$).

The normalization to one of $|\psi|^2$ implies the condition $\int |Y_l^m|^2 d\Omega = 1$ for the double integral calculated over the unit sphere. This is the origin of the normalization constant N_{lm}. The negative values of m are given by $Y_l^{-|m|}(\Omega) = (-1)^{|m|} Y_l^{|m|*}(\Omega)$.

The overall sign is $s = (-1)^m$ if $m \geq 0$ and $s = 1$ for $m < 0$, with the constant

$$N_{lm} = \left(\frac{2l+1}{4\pi} \frac{(l-|m|)!}{(l+|m|)!} \right)^{1/2}.$$

As seen in the expansion $v(x) = \sum_n a_n x^n$, the Legendre polynomials $P_l(x)$ (differentiated $|m|$ times with respect to x) only contain either even or odd powers of $\cos\theta$.

Hence the eigenfunctions $\psi(r)$ inherit a well defined parity, since they always meet the property: $\psi(-r) = \psi(r)$ (even parity) or $\psi(-r) = -\psi(r)$ (odd parity) $\forall r \in \mathbb{R}^3$.

There is a good physics reason for the above, since $\psi(r)$ actually represents a *stationary* solution to the Schrödinger equation with a central potential (as we shall see immediately), thus $|\psi|^2$ must have equal value at the opposite point ($r \to -r$), due to the spherical symmetry of the problem.

It is interesting noting that the parity of $\psi(\mathbf{r})$, as spherically symmetric function in $\mathbb{R}^3$ space, is determined by the body's *orbital motion*. In Classical Mechanics, it is not actually possible to talk about a well defined parity of a given trajectory (think about Kepler's elliptical orbits), and the whole idea of assigning a parity to a body's state of motion is *unique to Quantum Mechanics*. It is a useful exercise to show that the parity of ψ is exactly $(-1)^l$, being independent of the value of m.

The Legendre polynomials $P_l(x)$ can also be generated in a recursive way by means of the well known Rodrigues formula:

$$P_l(x) = \frac{1}{2^l} \frac{1}{l!} \frac{d^l}{dx^l}(x^2 - 1)^l,$$

and they are provided in Table 16.1 up to $l = 5$. In Table 16.2 we provide the detailed expression of the spherical harmonics up to $l = 3$, including the normalization constants.

A convenient way to show the $Y_l^m(\theta\phi)$ functions in 3D is to associate them with a surface such that the radial distance to the origin from a generic point on the surface of (θ, ϕ) spherical coordinates is proportional to $|Y_l^m(\theta\phi)|$. This provides a visual sensation of the probability distribution of electrons in atoms, molecules, etc. Their 2D projection onto the XZ or YZ planes already carries all the information, and these are called *polar diagrams*. We have represented both options in Figure 16.2, for the spherical harmonics with $l = 0$ and $l = 1$, in Figure 16.3 for $l = 2$ and in Figure 16.4 for $l = 3$, using *Wolfram Mathematica 11*. Note that the important information about the *phase* of Y_l^m is *not represented* in this kind of diagrams, which always show azimuthal symmetry, by construction.

A simpler representation is offered in Figure 16.5 for $l = 3$, where the quantum values of the opening angle between $\mathbf{L}$ and the Z-axis are explicitely shown, for all values of m.

The spherical harmonics form an orthonormal basis within the set of square-integrable complex functions defined on the unit sphere. This set can be understood as a subspace of the Hilbert space $L^2(\mathbb{R}^3)$, where the scalar product is defined as a double integral over the unit sphere. Accordingly, the spherical harmonics meet the orthonormality property:

$$\int Y_l^{m*}(\Omega) Y_{l'}^{m'}(\Omega) d\Omega = \delta_{ll'} \delta_{mm'},$$

wihere $d\Omega = d(\cos\theta) d\phi$ is the surface element of the 2D integral, with $\delta_{ab} = 0$ if $a \neq b$ and $\delta_{ab} = 1$ if $a = b$.

Chapter 16. THE ANGULAR MOMENTUM EIGENSTATES

Table 16.2: Spherical harmonics up to $l=3$.

$$Y_0^0 = \left(\tfrac{1}{4\pi}\right)^{1/2}$$

$$Y_1^0 = \left(\tfrac{3}{4\pi}\right)^{1/2}\cos\theta$$

$$Y_1^{\pm 1} = \mp\left(\tfrac{3}{8\pi}\right)^{1/2}\sin\theta\, e^{\pm i\phi}$$

$$Y_2^0 = \left(\tfrac{5}{16\pi}\right)^{1/2}(3\cos^2\theta - 1)$$

$$Y_2^{\pm 1} = \mp\left(\tfrac{15}{8\pi}\right)^{1/2}\sin\theta\cos\theta\, e^{\pm i\phi}$$

$$Y_2^{\pm 2} = \left(\tfrac{15}{32\pi}\right)^{1/2}\sin^2\theta\, e^{\pm 2i\phi}$$

$$Y_3^0 = \left(\tfrac{7}{16\pi}\right)^{1/2}(5\cos^3\theta - 3\cos\theta)$$

$$Y_3^{\pm 1} = \mp\left(\tfrac{21}{64\pi}\right)^{1/2}\sin\theta\,(5\cos^2\theta - 1)\, e^{\pm i\phi}$$

$$Y_3^{\pm 2} = \left(\tfrac{105}{32\pi}\right)^{1/2}\sin^2\theta\cos\theta\, e^{\pm 2i\phi}$$

$$Y_3^{\pm 3} = \mp\left(\tfrac{35}{64\pi}\right)^{1/2}\sin^3\theta\, e^{\pm 3i\phi}$$

The Dirac notation for quantum states is particularly suitable for angular momentum, in the form: $|\psi\rangle = |lm\rangle \equiv R(r)Y_l^m(\Omega)$. The mathematical problem of eigenstates we have analized may then be summarized in a succint way as follows:

$$L_z|lm\rangle = m\hbar|lm\rangle$$
$$L^2|lm\rangle = l(l+1)\hbar^2|lm\rangle \;,$$

with integer values ranging: $l = 0, 1, 2 \cdots \infty$, and $m = -l, \cdots, 0, \cdots +l$,

Complex functions $f(\Omega)$ defined on the unit sphere can be expanded in spherical harmonics: $|f(\Omega)\rangle = \sum_{lm} a_{lm}|lm\rangle$, with $a_{lm} = \langle lm|f(\Omega)\rangle$ and $\sum_{lm}|a_{lm}|^2 = 1$. The ensemble of states $\{|lm\rangle\}_{m=-l,\cdots,+l}$ generates a subspace of dimension $2l+1$ within the Hilbert space, that is formed by all functions $|f_l(\Omega)\rangle = \sum_m a_{lm}|lm\rangle$, with $\sum_m |a_{lm}|^2 = 1$.

17. THE RADIAL SCHRÖDINGER EQUATION

In a central force field with potential $U(r)$ only dependent on $r \equiv |r|$, angular momentum is conserved in Classical Mechanics. Let us see how the angular momentum eigenstates generate stationary solutions to the Schrödinger equation, thus offering a way to describe angular momentum conservation in Quantum Mechanics.

In Schrödinger's equation $\left(-\hbar^2 \Delta/(2m) + U(r)\right)\psi = E\psi$ we may use the form (15.4) of the Laplace operator in spherical coordinates. Upon substitution of the wave functions $\psi = R(r) Y_l^m(\theta, \phi)$ that by construction verify $L^2(Y_l^m) = l(l+1)\hbar^2 Y_l^m$, we easily arrive to the following ordinary differential equation (after cancelling Y_l^m from each term):

$$\frac{-\hbar^2}{2m}\left(\frac{\partial^2 R}{\partial r^2} + \frac{2}{r}\frac{\partial R}{\partial r}\right) - \left(E - U(r)\right)R + \frac{\hbar^2}{2m}\frac{l(l+1)}{r^2}R = 0. \qquad (17.1)$$

The centrifugal energy plays in this equation entirely the role of an additional radial potential (repulsive, due to the + sign, and strongly increasing for $r \to 0$). It may be absorbed in the definition of the so-called *effective potential*:

$$U_l(r) \equiv U(r) + \frac{l(l+1)\hbar^2}{2mr^2},$$

that reduces the equation to the form:

$$\frac{-\hbar^2}{2m}\left(\frac{\partial^2 R}{\partial r^2} + \frac{2}{r}\frac{\partial R}{\partial r}\right) + U_l(r)R = ER, \qquad (17.2)$$

known in the literature as radial Schrödinger's equation.

Still the equation is often written in a way that the function $u(r) \equiv rR(r)$ is taken as the unknown. A simple application of the chain rule allows it to be reinstated as:

$$\frac{-\hbar^2}{2m}u'' + U_l(r)u = Eu \ . \qquad (17.3)$$

Written in this way, the radial equation loses its first derivative, and takes the form of the Schrödinger equation in one dimension, formulated for $u(r)$ (the radial coordinate). The wave function can obviously be retrieved from the solution as $R(r) = u(r)/r$. However the boundary conditions are not the same: $u(r)$ is subject to the condition $u(0) = 0$, which is ensured by the regular character of ψ ($R(0) \neq \infty$), whereas no such condition exists for equation (17.2).

The function $u(r)$ has its own probabilistic interpretation, since $|u(r)|^2$ represents the so-called *radial probability density*:

$$\frac{dP}{dr} = r^2|R(r)|^2 = |u(r)|^2 \ ,$$

where dP is the probability to find the particle in the radial interval $(r, r+dr)$, with normalization $\int_0^\infty (dP/dr)dr = 1$. Note this interpretation is possible only because the spherical harmonics verify $\int |Y_l^m|^2 d\Omega = 1$, upon integration over the full solid angle, at any distance r.

Let us establish a key property of the radial Schrödinger equation: the wave function behaves close to the origin (for $r \to 0$) as $\psi \sim r^l$, i.e. it goes to zero faster the higher the l value (for $l > 0$). This property is conditional to the potential energy meeting the requirement $\lim_{r \to 0} r^2 |U(r)| = 0$, which means that the negative potential energy should not fall for $r \to 0$ more rapidly than the function $-1/r^2$.

In order to prove the asymptotic behaviour above, it is enough to neglect the potential energy $U(r)$ (and consequently, also E) as compared to the centrifugal energy in equation (17.3), by virtue of the previous consideration, and solve the so simplified differential equation in the limit of very small distances ($r \to 0$):

$$\frac{-\hbar^2}{2m}u'' + \frac{\hbar^2 l(l+1)}{2mr^2}u = 0 \ ,$$

which reduces to $u'' = l(l+1)u/r^2$. Seeking solutions of the form $u = Cr^a$ we obtain the following relationship, and possible values of a:

$$a(a-1) = l(l+1) \qquad \Longrightarrow \qquad a = l+1 \quad \text{or} \quad a = -l$$

The negative solution implies $\psi = cR(r) = cu(r)/r = cr^{-(l+1)}/r \sim r^{-l}$ which is divergent for $r \to 0$ and does not meet the $\mathbb{R}^3$ continuity requirement that the wave function must satisfy, thus the behaviour $\psi \sim r^l$ is proven.

The case $U(r) \sim -\beta/r^a$ with $\beta > 0$ and $a = 2$ is interesting, even if no concrete physical examples are known. Depending upon the value of l, all solutions for ψ are divergent or oscillating at the origin, and there is no ground state with finite and negative energy. Since the attraction force is stronger than the centrifugal force, the particle really "falls" to the origin and the wave function becomes singular. The same behaviour is found, of course, for $a > 2$ [1]. This phenomenon of falling to the origin equally occurs in Classical Mechanics, for the same potentials.

It is clear that a De Broglie plane wave of momentum $p = (0,0,p)$ cannot possess a well defined value of the modulus of the angular momentum $|L|$ (l), because the particle's position in the direction perpendicular to p is completely undefined (as a consequence of the uncertainty principle, since $\Delta p_x = \Delta p_y = 0$). This is yet another manifestation of the *ideal* character of the plane wave, as particles always emerge from some *finite* space region where they have been produced, and never show an infinite wave front. Given the discrete values of l, their impact parameter in collisions must therefore be subject to some kind of quantum footprint.

The following question is then raised: do quantum states of well defined angular momentum l exist in *free* motion? It is clear that the right tool to answer the question is the radial Schrödinger equation. Let us first analyze it for $l = 0$ (equation (17.3)): $-\hbar^2 u_0''/(2m) = E u_0$, or: $u''_0 = -k^2 u_0$, with $k = \sqrt{2mE}/\hbar$. Its general solution is:

$$R_0(r) = \frac{1}{r}\left(Ae^{ikr} + Be^{-ikr}\right).$$

The condition that $R_0(0)$ must be finite forces us to choose $A + B = 0$, thus:

$$R_0(r) = \frac{C_k}{kr}\sin kr,$$

where C_k defines a suitable normalization constant.

---

[1] for a detailed analysis, see "Quantum Mechanics I" A. Galindo and P. Pascual, Springer-Verlag, 1990, and "Quantum Mechanics" L. D. Landau and E. M. Liftshitz, Butterworth-Heinemann (1977).

Chapter 17. THE RADIAL SCHRÖDINGER EQUATION

Table 17.1: The spherical Bessel functions up to $l = 2$.

$$j_0(x) = \frac{\sin x}{x} \qquad j_1(x) = \frac{\sin x}{x^2} - \frac{\cos x}{x} \qquad j_2(x) = \left(\frac{3}{x^3} - \frac{1}{x}\right)\sin x - \frac{3}{x^2}\cos x$$

The above function is simple, and has been depicted in 2D in Figure 17.1(b), on the horizontal plane. To analize the case $l \neq 0$ we define $\rho \equiv kr$ as the function argument and introduce the new function $\chi_l(\rho) \equiv R_l(\rho)/\rho^l$. Then equation (17.1) can be rewritten as:

$$\chi''_l + \frac{2(l+1)}{\rho}\chi'_l + \chi_l = 0, \qquad (17.4)$$

of which we already know the solution for $l = 0$: $\chi_0 = R_0 = C_k \sin(\rho)/\rho$. To find all of them, we begin by proving the following inductive law: $\chi_{l+1}(\rho) = \chi'_l(\rho)/\rho$. For this purpose let us differentiate equation (17.4) with respect to ρ:

$$\chi'''_l + \frac{2(l+1)}{\rho}\chi''_l + \left(1 - \frac{2(l+1)}{\rho^2}\right)\chi'_l = 0,$$

and substitute therein the inductive law under test $\chi'_l = \chi_{l+1}\rho$, to obtain:

$$\chi''_{l+1} + \frac{2(l+2)}{\rho}\chi'_{l+1} + \chi_{l+1} = 0.$$

What we find is exactly the original equation (17.4), after the replacement of l by $l+1$, which proves the inductive law. Hence all solutions for $l \neq 0$ are obtained upon differentiation of our result for $l = 0$, according to the expression:

$$\chi_l = \left(\frac{1}{\rho}\frac{d}{d\rho}\right)^l \left(C_k \sin(\rho)/\rho\right).$$

But these are just the *spherical Bessel functions*, up to a phase factor $(-1)^l$, defined in the literature as:

$$j_l(x) \equiv (-1)^l \left(\frac{1}{x}\frac{d}{dx}\right)^l \frac{\sin x}{x},$$

which we provide in Table 17.1 up to $l = 2$. Thus we have shown that the solutions (finite at the origin) to the radial Schrödinger equation (17.3) for the *free* motion of a particle can be expressed as:

$$|Elm\rangle \equiv C_l\, j_l(kr)\, Y_l^m(\theta\phi). \qquad (17.5)$$

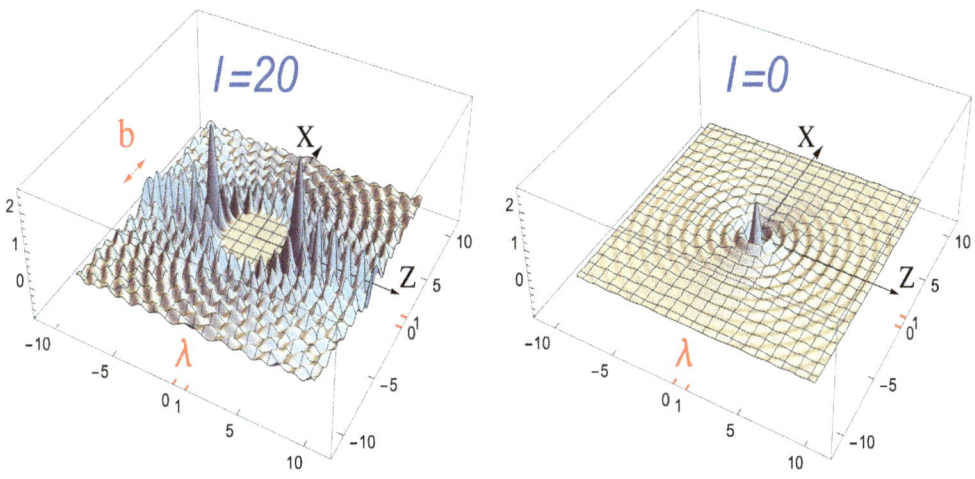

Figure 17.1: *Real wave function ψ_l of a free electron moving along the Z-axis with well defined energy E and $l = 20$, with $m_l = 0$ (left) and $l = 0$ (right). The function ψ_l has been evaluated in a generic plane that contains the Z-axis. The electron on the left rotates strongly about the vertical direction, whereas the one on the right does not rotate at all (S-wave). Note the emergence of the classical impact parameter b in the first case. Both functions have been represented on the same scale, just as they appear in the partial wave expansion of the plane wave e^{ikr} (17.7), including the $2l + 1$ factor. The De Broglie wavelength λ is also indicated, which defines the distance scale.*

We have denoted these solutions by the symbol $|Elm\rangle$, since they are quantum states characterized by their kinetic energy and by their angular momentum (note p is not well defined, but distributed over a spherical surface). The centrifugal force alone (in absence of an attractive potential) does not constrain the energy values, and the energy spectrum that is obtained is *continuous*: $E \in [0, +\infty)$.

The orthogonality of the above states in the Hilbert space for $E \neq E'$ needs to be expressed by means of a Dirac delta function, under the form:

$$\langle E'l'm'|Elm\rangle = \delta(E-E')\delta_{ll'}\delta_{mm'}, \qquad (17.6)$$

which allows to determine uniquely the normalization constant C_l. Note these states of the continuous spectrum, that include the plane waves themselves as a limiting case, are not square-integrable functions [2].

---

[2] for reference, the exact value of the constant C_l in (17.5) is: $C_l = (i^l/\hbar)\sqrt{2mk/\pi}$, see the book "Modern Quantum Mechanics" J. J. Sakurai, Pearson (2014), p. 398, for a detailed derivation.

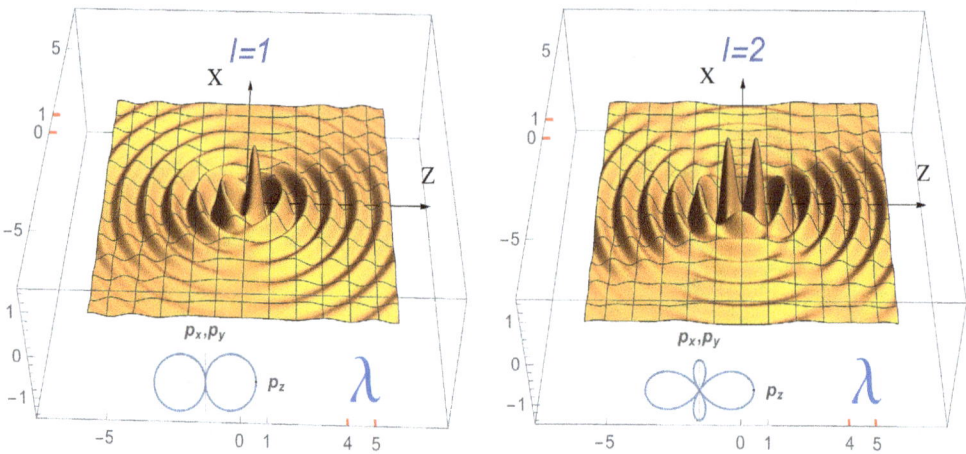

Figure 17.2: The $l = 1$ (P-wave) and $l = 2$ (D-wave) components of the expansion of e^{ikr}: $i^l(2l+1)P_l(\cos\theta)j_l(kr)$, evaluated on the horizontal plane. For $l = 1$ the i factor has been omitted. Note in this case the odd parity of the wave (the opposite sign minima are hidden). These are also states of well defined kinetic energy $E = \hbar^2 k^2/(2m)$ and angular momentum l, with $m_l = 0$. Because the distance scale is refered to the De Broglie wavelength, the image remains the same, irrespective of the energy and particle mass. The polar diagrams of their Fourier transforms (17.8) (spherical harmonics) are also shown below.

Bearing in mind the above idea, we may now improve our mathematical description of the De Broglie plane wave e^{ikr}, in the direction perpendicular to the $\boldsymbol{k}$ vector. It is clear that the wave can be expressed as a series expansion over the $|Elm\rangle$ set of orthonormal states, for a given value of the energy: $E = \hbar^2 k^2/(2m)$.

Before we write down the expansion, we must give thought to the role played by the quantum number m in expression (17.5). Let us recall that, once l is fixed, m indicates the angle between the $\boldsymbol{L}$ vector (rotation axis) and the Z-axis of the spherical coordinates. Yet in the absence of a magnetic field $\boldsymbol{B}$ or other external reference in our Hamiltonian, the orientation of the Z-axis is completely arbitrary. It may then be aligned with the momentum of the particle, as defined by the $\boldsymbol{k}$ vector. Thus we have $k_x = k_y = 0$, and the rotation is perpendicular to $\boldsymbol{k}$ ($L_z = 0$), hence $m = 0$. Let us proceed in this way and see whether a representation of the plane wave is possible, using only spherical harmonics with $m = 0$. Thus we must take the spherical harmonic to be: $Y_l^0 = \sqrt{(2l+1)/4\pi}\, P_l(\cos\theta)$, according to (16.6).

The intended expansion of the plane wave can indeed be achieved, as stated below [3], over the previously discussed orthonormal base of the Hilbert space:

$$e^{ikr} = e^{ikr\cos\theta} = \sum_{l=0}^{\infty} i^l (2l+1) P_l(\cos\theta) j_l(kr) \,. \qquad (17.7)$$

The azimuthal symmetry (independence of ϕ) of the above expression about the k vector is clear, just because $m = 0$. Note that the functions $j_l(x)$ and $P_l(x)$ are both *real*, so that the phase of the plane wave in the left-hand side of equation (17.7) is built by the powers:

$$i^l = 1, i, -1, -i, 1, i, \cdots \quad l = 0, 1, 2, 3, 4, 5 \cdots \infty \,.$$

More specifically, its real part is built by the even terms of the series, while the imaginary part is built by the odd terms, since only phases that are multiples of $\pi/2$ are actually present.

Notice that the *direction* of propagation $\pm k$ of the De Broglie wave (recall Figure 7.1) is encoded by a factor $(-1)^l$ when taking the *complex conjugate* of expression (17.7). This factor is equivalent to a change of sign of all the Legendre polynomials according to their parity $(-1)^l$.

Expression (17.7) is called *partial wave expansion* in the literature, and it plays a key role in the theory of particle scattering by a central force field, where it helps to describe the components of both the incident wave and each of the scattered outgoing waves. These components intercept increasing areas of the wave front perpendicular to the k vector. As an illustration, we have represented in Figure 17.1 two particular components of the partial wave expansion of a particle, with $l = 20$ and with $l = 0$.

As already mentioned, the transverse coordinate x is delocalized, even if $|L|$ is well defined. In Classical Mechanics we talk about the *impact parameter* $b = x$, assuming that this coordinate is itself well defined. Such degree of definition in the partial waves is indeed better and better, as we approach the classical limit $l \to \infty$. We have: $b \sim l/k = \lambdabar$, because $|L| = rp\sin\theta = b\hbar k = \sqrt{l(l+1)}\hbar \sim \hbar l$, $\lambdabar \equiv \lambda/(2\pi)$ being the reduced De Broglie wavelength.

---

[3] the interested student can easily verify, taking into account the orthogonality of the Legendre polynomials under the form $\int_{-1}^{+1} P_l(x) P_{l'}(x) dx = \delta_{ll'} 2/(2l+1)$, that formula (17.7) is readily deduced from the following representation of the spherical Bessel functions, due to Poisson: $j_l(kr) = \frac{1}{2i^l} \int_{-1}^{+1} e^{ikr\cos\theta} P_l(\cos\theta) d(\cos\theta)$, and well documented in the literature. Just multiply both sides of (17.7) by $P_{l'}(\cos\theta)$ and integrate over $(-1, 1)$.

Thus for $\theta = \pi/2$, the dimensionless argument of the Bessel function kr runs over a continuous version of the l values. Indeed the $j_l(x)$ function, for large values of l ($l \gg 1$), is nearly zero for $x \lesssim l$, and suddenly increases around that point, before going into oscillations, as it can be seen in Figure 17.1 (left), where the emergence of the impact parameter b in the quantum mechanical description can be noticed, for $l \to \infty$.

We have expressed the states $|Elm\rangle$ in the coordinate space, but we could equally express them in the momentum space. The result is provided below, with $\hat{k} \equiv \mathbf{k}/k$:

$$\langle \mathbf{k}|Elm\rangle \equiv \frac{1}{(2\pi)^{3/2}} \langle e^{i\mathbf{k}\mathbf{r}}|Elm\rangle = \delta\left(E - \frac{\hbar^2 k^2}{2m}\right) B_k Y_l^m(\hat{k}), \qquad (17.8)$$

which is nothing but the Fourier transform [4] of expression (17.5).

The previous expression is entirely natural, since the spherical harmonics tell us the amplitude for the particle to be located at a given point on the unit sphere that, in this case, corresponds to a given orientation of its momentum. The Dirac delta function comes from the normalization of the $|Elm\rangle$ states of the continuum spectrum defined by equation (17.6).

As a particular case of (17.8), it can be shown from the constant values B_k and C_l, and recasting $\delta\left(E - \frac{\hbar^2 k_0^2}{2m}\right) = \left(\frac{m}{\hbar^2 k_0}\right)\delta(k - k_0)$, that the Fourier transform of the plane wave (17.7), with wavenumber k_0, is expressed as:

$$f(\mathbf{k}) = \sum_l A(k)(2l+1)P_l(\cos\theta_k) \qquad A(k) = \sqrt{\frac{\pi}{2}} \frac{1}{k_0^2} \delta(k - k_0).$$

It is worth noting that if the summation of the series does *not* proceed to infinity, then the wave is no longer *forward* directed (along $\mathbf{k}_0$), and the particle may point in *all* directions $\hat{k}$, including backwards, for low l values.

In Figure 17.2 we have calculated, using *Wolfram Mathematica 11*, the partial waves for $l = 1$ (P-wave) and for $l = 2$ (D-wave), including their representation in the momentum space, which, as spherical harmonics, are best represented by polar diagrams on the same coordinate axes.

---

[4] an indirect proof of expression (17.8) can be attained by checking that it is precisely the one needed to recover the partial wave expansion (17.7). Indeed, starting from $|e^{i\mathbf{k}\mathbf{r}}\rangle = \int \sum_{lm} \langle Elm|e^{i\mathbf{k}\mathbf{r}}\rangle |Elm\rangle dE$ and substituting the above expression therein, taking into account the very useful property of the spherical harmonics: $P_l(\cos\theta) = P_l(\hat{k}\hat{r}) = \left(\frac{4\pi}{2l+1}\right)\sum_{m=-l}^{l} Y_l^m(\hat{k})Y_l^{m*}(\hat{r})$, one arrives to (17.7) upon integration of the Dirac delta function, after recalling (17.5). The exact value of the constant is $B_k = \hbar/\sqrt{mk}$.

18. THE BOHR MAGNETON

Every charged particle with non zero angular momentum (either from orbital or spin rotation), necessarily acquires a magnetic moment μ. For an electron with charge $-|e|$ and mass m_e, that ocupies the point r with momentum $p = m_e v$ (here non relativistic), we know precisely at any time t the magnetic moment associated to its orbital motion:

$$\mu = \frac{1}{2}(r \times ev) = \left(\frac{e}{2m_e}\right) L .$$

The above expression is well known in electromagnetism [1], and we shall not go here into its basics (the factor $1/2$ arises from using the cross product). Note the electron *is not required to run over a differentiable trajectory at all*. What is stated by electromagnetism is that, concerning orbital motion, the vectors μ and L are proportional to each other at any time. The proportionality factor is precisely established, subject to the sign of the electric charge $\pm e$ (parallel or antiparallel).

The magnetic moment is measured in Am^2 (Amperes meter squared) or in JT^{-1} (Joules per Tesla), that are equivalent units in the International System (SI) [2]. The factor $\omega_0 \equiv |e|/(2m_e)$ is called *standard gyromagnetic ratio*, and takes the value $\omega_0/(2\pi) = \nu_0 = 13996\, MHz\, T^{-1}$ for an electron. Note that for a proton, the gyromagnetic ratio is approximately 2000 times smaller.

---

[1] see for example the book by D. J. Griffiths "Introduction to Electrodynamics", Pearson (2013).
[2] we strongly discourage the use of obsolete unit systems that may hamper a fluid interaction between the fields of physics and engineering.

Chapter 18. THE BOHR MAGNETON

All ideas of the preceding chapters concerning the uncertainty principle, the impossibility to measure independently, with infinite precision, the three components of L, and the quantum values of its projection along any axis Z, are intrinsically applied now to the μ vector. The magnetic moment is manifest when an external magnetic field B is applied to the particle, that acquires the extra energy:

$$E = -\mu B \ .$$

The natural axis onto which the μ vector is projected to take its quantum values is the B-field direction. The existence of a nonzero minimum value of $|L_z|$ ($\hbar$), translates itself into a *quantum of magnetic moment* for the electron, that is called Bohr magneton in the literature:

$$\mu_B \equiv \frac{|e|\hbar}{2m_e} = 9.274 \times 10^{-24} JT^{-1} = 5.788 \times 10^{-5} eVT^{-1} \ .$$

We may now write the following generic vectorial equation (for the electron, with negative charge), holding both for the physical magnitudes and for the operators that represent them in Quantum Mechanics:

$$\mu_l = -\frac{\mu_B}{\hbar} L \ ,$$

equation which does not actually depend on $\hbar$, but it is standard to write in this way. Accordingly we have, for $l = 0, 1, \cdots \infty$, and arbitrary orientation of the Z-axis:

$$|\mu_l| = \sqrt{l(l+1)}\mu_B \qquad \mu_{l,z} = -m_l \mu_B \qquad m_l = -l, -l+1, \cdots, 0, \cdots +l \ .$$

If the electron wave function ψ is known, the mean value of the above operators should be calculated, and the following equation will be met $\forall \psi \in L^2(\mathbb{R}^3)$:

$$\langle \mu_l \rangle = -\frac{\mu_B}{\hbar} \langle L \rangle \ .$$

Now the question comes what is the correct gyromagnetic ratio for the electron spin. We must bear in mind that a proper description of spin requires a *model* about the nature of what does it actually rotate within a given particle. The model may simply mean the specification of its quantum values, in the context of a relativistic theory [3].

---

[3] that is the case, for example, of the Dirac theory for the relativistic electron, that emerges from the *Dirac equation*.

Moreover *only experiment* can validate the quantum description of the spin of a given particle. In the case of the electron, the experiment was performed in 1922 by Stern and Gerlach, and will be dealt with in Chapter 21. It was a historical experiment in Physics, and it meant at the time a challenge to the consolidation of Quantum Mechanics, as we shall see.

We already anticipate here that the results of the Stern and Gerlach experiment are in apparent contradiction with the postulates of Quantum Mechanics, as we have applied them to angular momentum in Chapter 16. We will nevertheless write down below the most general equations that still allow compatibility of the experiment with the postulates, but in a somehow *broader view* of Quantum Mechanics.

First of all, let us admit that the gyromagnetic factor may differ from the standard one $|e|/(2m_e)$, and parametrize the deviation with a factor g_s. Secondly, let us accept that the spin S (intrinsic angular momentum) may be quantized in a slightly more general way than orbital angular momentum, for a given orientation of the Z-axis, such that:

$$|S| = \sqrt{s(s+1)}\hbar \qquad S_z = m_s\hbar \qquad m_s = -s, -s+1, \cdots +s \,,$$

but where the s value is not necessarily a *positive integer*, being allowed to be a *half-integer*. Note that m_s does increase by one-unit steps. If s is non integer, we are indeed in contradiction with our earlier results of Chapter 16, based on the probabilistic interpretation of the wave function ψ in Quantum Mechanics (implying its single-valued character in $\mathbb{R}^3$) and on the assessment of the associated Legendre equation.

Should we allow non integer values of s to accommodate the experimental result, as will be seen in Chapter 21, and considering not only the electron, but every possible particle, the *intrinsic magnetic moment* (associated with the spin) must be written as:

$$\mu_s = g_s \frac{\mu_s}{\hbar} S \qquad |\mu_s| = g_s \sqrt{s(s+1)} \mu_s \qquad \mu_{s,z} = g_s m_s \mu_s \,,$$

where $\mu_s \equiv \mu_B$ for the electron and $\mu_s \equiv |e|\hbar/(2m_p) \equiv \mu_N$, defined by convention as the nuclear magneton, for both proton and neutron. For every particle, we must let the experiment determine the exact values of s and g_s (the latter being signed, even for a neutral particle), to constrain whatever hypotheses are made concerning its internal structure, or the lack of it thereof.

Let us now come back to equation $E = -\mu B$, the energy acquired by a magnetic dipole in an external magnetic field, that is generically valid whether μ originates from orbital motion (μ_l) or spin (μ_s). Having defined μ as an *operator* in the Hilbert space, for *orbital* motion, the above equation must be understood as the expression for the *Hamiltonian of the interaction* of the particle with the *B*-field. Thus for an electron in orbital motion:

$$H_L = \frac{\mu_B B}{\hbar} L_z \,.$$

More generally, we have: $H_L = -(q/(2m))BL_z$, with q/m being the particle's charge/mass ratio. For example, $H_L = -(\mu_N B/\hbar)L_z$ for a proton. It is assumed that the *B*-field is high enough that its quantum fluctuations can be neglected, that is, lying on a well defined axis: $B = (0,0,B_z)$.

Such Hamiltonian may be used, as part of Schrödinger's equation, to analize the time evolution of the wave function ψ of the particle under the external field. Its eigenvalues are obvious, given the eigenvalues of the L_z operator. Notice that both H_L and L_z take the form of an *Hermitian matrix* operating in the subspace of quantum states associated with a given value of l (dimension $2l+1$).

For the spin case the Hamiltonian $H_S = -g_s(\mu_s/\hbar)BS_z$ is analogously defined, taking the form of a Hermitian matrix that operates on a subspace of dimension $2s+1$. Its eigenvalues are also obvious from those of the S_z operator. Its matrix form will be seen in Chapter 21. We have $\mu_s \equiv \mu_B$ for the electron, or $\mu_s \equiv \mu_N$ (nuclear magneton) for proton and neutron, with measurable *signed* g_s factors, that are different for each of the 3 cases.

For an electron in an atom or molecule, both H_L and H_S mean an *additional* term to the Hamiltonian of the system. It has the effect of *splitting* the emitted photon energy lines, that is known as Zeeman effect in the literature. Pieter Zeeman was the Dutch physicist who first studied the effect of magnetic fields on light sources, crucially important for the early developments of Quantum Mechanics.

For example, an atomic transition between a *p*-orbital that decays to an *s*-orbital will be split into two additional frequency lines, corresponding to the variation of the upper energy by $\Delta E = 0, \pm \mu_B B$. It can easily be checked numerically that, even for field values as high as several Tesla, the frequency splitting is *small* in relative terms (note $E \approx -1\,\text{eV}$). The S_z operator will be seen in Chapter 21, and equally raises Zeeman effect.

19. THE LINEAR POLARIZATION STATES

According to what was seen at the end of Chapter 16, the most general state of orbital rotation of a particle is defined by a linear combination of eigenstates of the angular momentum, with *complex* coefficients. Indeed, simultaneous rotation about the Z-axis in two opposite directions is allowed by the Feynman propagator, and fully consistent with the principles of Quantum Mechanics. Let us see in detail the structure and physical meaning of such states, that are involved in so many cases of physical interest.

Before doing so, let us consider the following identity between complex numbers, easily derived $\forall \phi, \alpha \in [0, 2\pi]$ [1]:

$$\frac{1}{2}\left(e^{i\phi} + e^{i\alpha}e^{-i\phi}\right) = \cos\left(\phi - \frac{\alpha}{2}\right)e^{i\alpha/2} . \qquad (19.1)$$

There is no mistery in this equation: the superposition of two opposite circular motions on the unit circle ($e^{\pm i\phi}$) generates a straight line that periodically goes through *zero* when running the phase ϕ. The inclination of the straight line is governed by the phase shift α between both complex numbers. It is broadly used, from mechanics to electromagnetism.

For example, it allows us to describe precisely the linear polarization of light, when the above sum represents the electric field vector E of an electromagnetic wave with frequency ω, the phase being $\phi = \omega t$:

$$E(t) = \frac{1}{2}\left(e^{i\omega t} + e^{i\alpha}e^{-i\omega t}\right) E_0 = \cos\left(\omega t - \frac{\alpha}{2}\right)e^{i\alpha/2} E_0 .$$

---

[1] the analogous identity: $\frac{1}{2i}\left(e^{i\phi} - e^{i\alpha}e^{-i\phi}\right) = \sin\left(\phi - \frac{\alpha}{2}\right)e^{i\alpha/2}$ can also be derived in a similar way.

Though we will soon come back to the photon of this wave, let us first apply the above identity to the quantum superposition of two rotational states of an electron. Within the Hilbert subspace with $l=1$ (dimension 3, with orthonormal base $|m_l\rangle$, $m_l = -1, 0, +1$), we define the `linear polarization states` of the electron as:

$$|\psi_\alpha\rangle \equiv |\text{lin}, \alpha/2\rangle \equiv \frac{1}{\sqrt{2}}|+1\rangle + \frac{e^{i\alpha}}{\sqrt{2}}|-1\rangle. \tag{19.2}$$

Using the expression of the spherical harmonics $Y_1^{\pm 1}$ (Table 16.2), and the analogous identity to (19.1), we can write the angular part of the above wave function as:

$$\psi_\alpha(\theta, \phi) = -i\sqrt{\frac{3}{4\pi}} \sin\theta \sin(\phi - \alpha/2) e^{+i\alpha/2}. \tag{19.3}$$

Its radial part is actually determined by the force field imprinted on the electron, with: $\psi = f(r)\psi_\alpha(\theta, \phi)$, and $\int_0^{+\infty} r^2 |f(r)|^2 dr = 1$, that defines the spatial extension of the wave function (for instance, in an atom or molecule, the size will be of the order of the Bohr radius, a_0). In Figure 19.1 the function $|\psi_\alpha(\theta, \phi)|$ has been depicted for two particular values $\alpha = \pi$ ($|X\rangle$) and $\alpha = 0$ ($|Y\rangle$), along with each state $|\pm 1\rangle$ separately.

It is seen how the effect of the linear superposition breaks up the azimuthal symmetry observed on the spherical harmonics $|\pm 1\rangle$. As the electron simultaneously rotates in opposite directions about the Z-axis, the probability density clusters at a given orientation of the azimuth ϕ, where $|\psi_\alpha|^2$ is maximal, defined by $\phi = \alpha/2 + \pi/2$. Note the factor 2 that arises between the phase shift α and the linear polarization angle $\alpha/2$.

For $\alpha = \pi$ and $\alpha = 0$, these orbitals are called p_x and p_y in Chemistry, for reasons that are apparent from Figure 19.1. The p_z orbital corresponds to the state $|0\rangle$, with its probability density oriented toward the Z-axis (recall Figure 16.2). The importance of the spatial orientation of these states for the chemical bonds of Carbon, Oxygen, etc, is well known.

A physical definition of the X (Y) axes is needed, in the Hamiltonian of the system, to bring the electron to these p_x (p_y) states. An example is given by the (unhybridized) p_x and p_y orbitals of two Carbon atoms, when they bind together to form the Ethyne (C_2H_2) molecule. The p_x and p_y orbitals of each atom, that are orthogonal states forming a 90° angle, overlap with those in front to contribute with two bonds between the Carbon atoms (the so-called π-bonds). The Z-axis is defined in this case by the collinear *sp* (hybrid) orbitals that form the third bond.

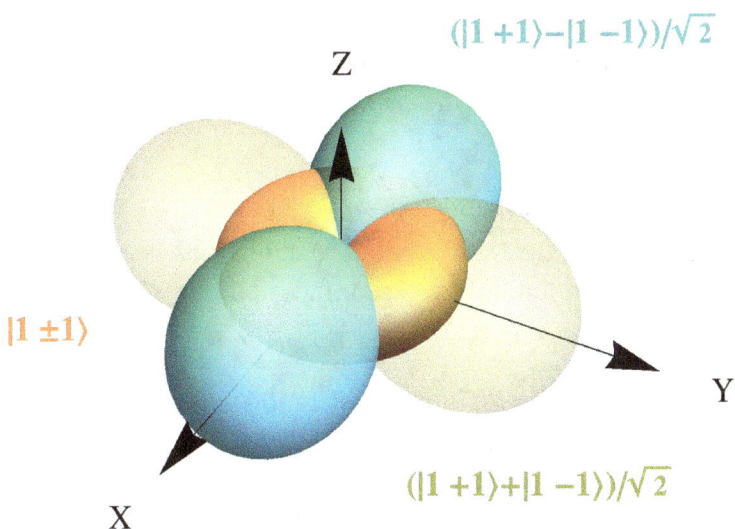

Figure 19.1: *Modulus of the angular wave function of an electron in spherical coordinates $|\psi_\alpha(\theta,\phi)|$, as defined by expression (19.3), for $\alpha = \pi$ (blue), and $\alpha = 0$ (light green). Each of the superposed states $|\pm 1\rangle$ are depicted in orange, being equal to each other. The distance from the surface's points to the origin of coordinates is proportional to the angular probability density to find the electron with that particular orientation (θ,ϕ). The relative scale among all functions has been rigorously kept. The same drawings also represent, independently, the polarization states of an individual photon that propagates along the positive Z-axis, linearly polarized along the X-axis (blue), along the Y-axis (green), or circularly polarized (orange). In the photon case the probability density refers to the orientation of the electric field $\mathbf{E}$.*

An interesting case takes place when, in a linear polarization state, the superposed states in $|\psi_\alpha\rangle$ acquire different energies $E_{1,2} = \pm\mu_B B$, due to an external B-field on the Z-axis. Then the position of the maximum probability density in the azimuth, defined by $(\alpha + \pi)/2$ in equation (19.3), and represented in Figure 19.1 (with either p_x or p_y) gets a rotary motion: $\alpha(t)/2 = \alpha_0/2 + \omega_L t$, with α_0 being its initial phase. Its angular frequency: $\omega_L = \mu_B B/\hbar = (E_1 - E_2)/(2\hbar)$ is well defined. Indeed, each state gets a frequency: $\omega_{1,2} = \pm\mu_B B/\hbar$, when evolving with Schrödinger's equation $\psi_\alpha(t) = \mathscr{U}(t)\psi_\alpha(0)$, such that the state can be expressed as:

$$|\psi_\alpha(t)\rangle = \frac{e^{-i\omega_1 t}}{\sqrt{2}}\left(|+1\rangle + e^{i(\alpha_0 + 2\omega_L t)}|-1\rangle\right). \qquad (19.4)$$

This time evolution provides a quantum interpretation of the phenomenon known in classical electromagnetism as *Larmor precession*, by which a magnetic dipole μ precesses about the external magnetic field B with frequency proportional to the product of their magnitudes, the Larmor frequency [2].

The quantum states defined by equation (19.2) have another important application, beyond the framework of rotational states of a particle of mass $m \neq 0$ (electron, proton, neutron, etc). Indeed they also represent individual photon polarization states.

To understand this, take into consideration that in a *circularly polarized* electromagnetic wave, the electric field E and the magnetic field B both rotate in the plane perpendicular to the direction of propagation of the wave, either clock-wise, or anti-clock wise. Thus each individual photon must carry that information. The electromagnetic wave is a *vector* field, whose state of rotation is associated with the photon spin.

Not only the *quantum* carries energy $E = \hbar \omega$ and momentum $p = \hbar \omega / c$, but also *angular momentum*, as the wave itself does. An intrinsic photon angular momentum of $\pm \hbar$ was measured in 1931 by the Indian physicists Chandrasekhara Raman and Suri Bhagavantam [3] who showed that the degree of depolarization of the light scattered by O_2, CO_2 and *NO* molecules was in excellent agreement with the above quantitative hypothesis.

The observed value $S_z = \pm \hbar$ (Z-axis along the photon momentum) suggests that $s = 1$ (the modulus of the spin) in our earlier discussion of Chapter 18. Because s is an *integer* value, it is a natural fit for the quantum theory. However, it remains to be explained why the experiments never observe the value $S_z = 0$.

---

[2] the Larmor precession must in addition take into account the electron spin, that we shall see later on. Note that in the above state $|\psi_\alpha\rangle$, $\langle \mu \rangle = 0$ is met and the quantum precession actually takes place in *two* opposite and superposed dipoles. Nevertheless the displacement seen of the electron density is perfectly representative of this phenomenon.

[3] C. V. Raman and S. Bhangavantam, Indian J. Phys., Vol 6, p. 353-366 (1931). For a modern version, including rotation induced to an optomechanical device, see the article L. He, H. Li, M. Li Sci. Adv. 2016; 2:e1600485, September 2016, 10.1126/sciadv.1600485.

The explanation is conceptually simple, and it is found in the *relativistic* character of the electromagnetic wave, that propagates at light velocity c. If the electric field E could rotate in a plane that contains the Z-axis, and thus vibrate along the Z-direction, longitudinal displacements could be observed e.g. of electrons, pushed and pulled by the wave. But this is against the relativistic *Lorentz contraction* to zero that any length undergoes at light velocity. Hence the non observation of $S_z = 0$ is explained by the *zero mass* of the photon, i.e. by its propagation at light velocity.

Thus the light polarization modes actually mean internal quantum states of every photon, that live in a Hilbert space of dimension two. The orthonormal base elements $|S_z\rangle$ can be denoted by the symbols: $|+1\rangle$ and $|-1\rangle$, or also $|R\rangle$ and $|L\rangle$ (right-handed photon and left-handed photon, respectively). Note that both of these states have negative parity, being associated with $l = 1$ spherical harmonics. Therefore, the photon is a quantum state of `negative intrinsic parity`.

There are as many light polarization states as there are *complex* linear combinations $\chi = c_1|+1\rangle + c_2|-1\rangle$ of the above states, subject to the normalization $|c_1|^2 + |c_2|^2 = 1$. Given that the overall phase is meaningless for an isolated photon, these are precisely the polarization states $|\text{lin}, \alpha/2\rangle$ defined by expression (19.2), when $|c_1| = |c_2|$. The meaning of the phase difference α is now clear: it governs (divided by 2) the vibration angle of the electric field with respect to the horizontal plane, in the linearly polarized light. For $|c_1| \neq |c_2|$ we get the elliptical polarization states.

In consequence, the states depicted in Figure 19.1 also represent the wave function of an *individual photon* propagating along the Z-axis, with either linear or circular polarization. The distance from the surface's points to the origin of coordinates must be understood as proportional to the probability density that the electric field is oriented in that particular direction (strictly speaking, the potential vector field A defines the angle). Keep in mind that the electric field and the magnetic field are both subject to the uncertainty principle.

20. THE HYDROGEN ATOM

Let us solve the radial Schrödinger equation for the Coulomb potential, where an electron is subject to the attractive potential of Z protons at a distance r:

$$U(r) = \frac{-Ze^2}{4\pi\varepsilon_0}\frac{1}{r}.$$

We shall find the energy spectrum we already know from Chapter 2.3, and thus confirm that the semiclassical quantization actually provides the exact result. Of course, the analysis with Schrödinger's equation will allow us to know in addition the wave functions and to determine their angular momentum and quantum degeneracy.

We leave aside the case with $E > 0$, for which we may find certain conceptual guide from our discussion at the end of Chapter 17 (when the field is very weak and the electron is nearly free), and focus on the case $E < 0$ that corresponds to the bound states (Kepler's orbits, in Classical Mechanics).

Being a central potential, we must seek solutions in spherical coordinates to the radial equation $H\psi = E\psi$, of the type:

$$\psi(r\theta\phi) = R_l(r)Y_l^{m_l}(\theta\phi).$$

We encode the energy value and the interaction coupling in the two constants:

$$\beta \equiv \frac{\sqrt{-2mE}}{\hbar} > 0 \qquad \gamma \equiv \frac{2mZe^2}{4\pi\varepsilon_0\hbar^2},$$

Chapter 20. THE HYDROGEN ATOM

Table 20.1: Laguerre polynomials up to $q = 5$

$L_0 = 1$	$L_1 = -x + 1$
$L_2 = \frac{1}{2}(x^2 - 4x + 2)$	$L_3 = \frac{1}{6}(-x^3 + 9x^2 - 18x + 6)$
$L_4 = \frac{1}{24}(x^4 - 16x^3 + 72x^2 - 96x + 24)$	$L_5 = \frac{1}{120}(-x^5 + 25x^4 - 200x^3 + 600x^2 - 600x + 120)$

Table 20.2: Radial functions R_{nl} for Hydrogen up to $n = 2$

$$R_{10} = 2\left(\frac{Z}{a_0}\right)^{3/2} e^{-Zr/a_0} \qquad R_{21} = \frac{1}{\sqrt{3}}\left(\frac{Z}{2a_0}\right)^{3/2}\left(\frac{Zr}{a_0}\right) e^{-Zr/2a_0}$$

$$R_{20} = 2\left(\frac{Z}{2a_0}\right)^{3/2}\left(1 - \frac{Zr}{2a_0}\right) e^{-Zr/2a_0}$$

so that the radial Schrödinger equation (17.1) takes the form:

$$R_l'' + \frac{2}{r}R_l' + \left(\frac{\gamma}{r} - \beta^2 - \frac{l(l+1)}{r^2}\right)R_l = 0.$$

Without loss of generality, it is convenient to perform a change of function from $R_l(r)$ to $\xi_l(r)$, in order to allow a precise assessment of the solutions that are square-integrable and those which are not:

$$R_l(r) \equiv \xi_l(r) e^{-\beta r}.$$

On application of the chain rule, the following differential equation for the function $\xi_l(r)$, already simplified, is obtained:

$$\xi_l'' - 2\left(\beta - \frac{1}{r}\right)\xi_l' + \left(\frac{\gamma - 2\beta}{r} - \frac{l(l+1)}{r^2}\right)\xi_l = 0.$$

We now investigate the asymptotic behaviour for $r \to \infty$ of the solutions to the above equation. Should the product $\xi_l(r)e^{-\beta r}$ not go to zero at infinity, the wave function ψ would not be square-integrable, thus being unacceptable as a bound state in Quantum Mechanics.

Table 20.3: Radial functions R_{nl} for Hydrogen for $n=3$

$$R_{32} = \frac{2\sqrt{2}}{27\sqrt{5}} \left(\frac{Z}{3a_0}\right)^{3/2} \left(\frac{Zr}{a_0}\right)^2 e^{-Zr/3a_0}$$

$$R_{31} = \frac{4\sqrt{2}}{9} \left(\frac{Z}{3a_0}\right)^{3/2} \left(\frac{Zr}{a_0}\right) \left(1 - \frac{Zr}{6a_0}\right) e^{-Zr/3a_0}$$

$$R_{30} = 2 \left(\frac{Z}{3a_0}\right)^{3/2} \left(1 - \frac{2Zr}{3a_0} + \frac{2(Zr)^2}{27a_0^2}\right) e^{-Zr/3a_0}$$

Let us write down the power series expansion of $\xi_l(r)$ in $r \in [0, +\infty)$, and establish the recursive law that is derived for its coefficients on application of the differential equation. Prior to that, we take into account the result obtained in Chapter 17, showing the asymptotic behaviour of the solutions to the radial Schrödinger equation for $r \to 0$ as r^l, due to the centrifugal barrier. According to it, the series expansion should be written as:

$$\xi_l(r) = \sum_{k=l}^{\infty} a_k r^k = a_l r^l + a_{l+1} r^{l+1} + \cdots$$

since terms with $k < l$ cannot contribute. To establish the recursive law we need to differentiate the above expression up to two times, and then group the terms with equal powers of r. It is a bit laborious, and we do not show it here in full detail. Just point out though that ξ_l'', $(2/r)\xi_l'$ and $(-l(l+1)/r^2)\xi_l$ are coupled to the same powers: r^{k-2}. The final result for the recursive law is then obtained:

$$a_k = \frac{2k\beta - \gamma}{k(k+1) - l(l+1)} a_{k-1} \qquad k \geq l+1 \ . \qquad (20.1)$$

At this point we make use of the Lemma of Asymptotic Similarity stated at the beginning of Chapter 16, already used in a similar way in the context of the associated Legendre equation. The goal is to assess the asymptotic behaviour of the solution at the boundary of the radius of convergence, in this case $r \to +\infty$, by comparing its recursive law with that of a known test function. The test function this time is a simple exponential:

$$\frac{1}{r} e^{2\beta r} = \frac{1}{r}(1 + 2\beta r + \cdots) = \frac{1}{r} \sum_{k=0}^{\infty} \frac{(2\beta r)^k}{k!} \ ,$$

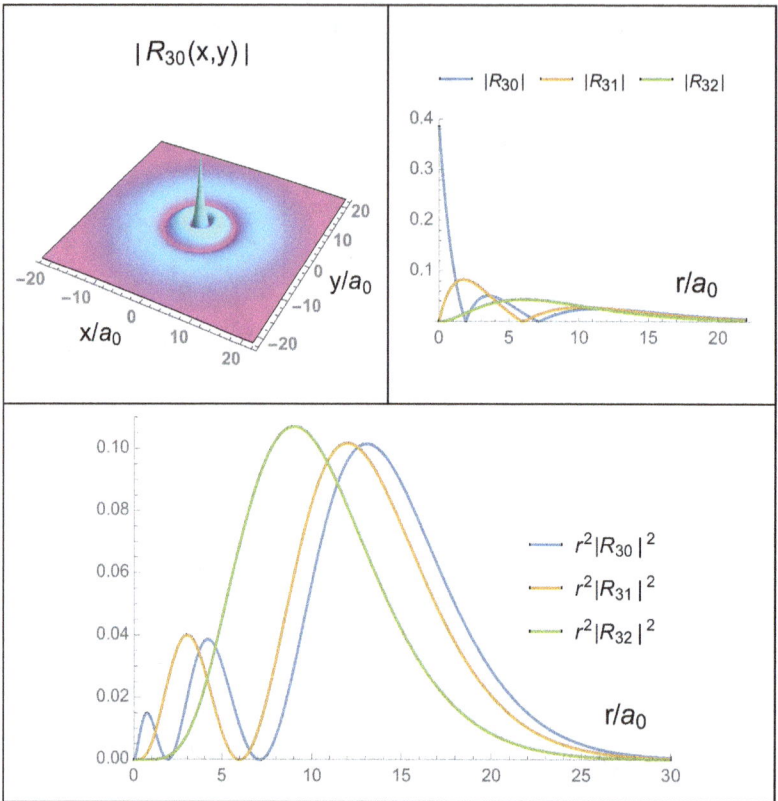

Figure 20.1: *Top left: probability density in the horizontal plane for the 3s state of Hydrogen ($n = 3, l = 0$), the distance being measured in units of the Bohr radius. Top right: comparison of the modulus of the radial wave function for the three states of Hydrogen with $n = 3$: $l = 0$ (previously represented), and $l = 1, 2$. Bottom: radial probability density of the three states above, where the effect of the centrifugal barrier on the electron is clearly observed.*

whose recurrence law is, straightforwardly

$$\frac{a_k}{a_{k-1}} = \frac{2\beta}{k},$$

that exactly coincides with that expressed in equation (20.1) for $k \to \infty$. As the exponential function is divergent for $r \to +\infty$ for $\beta > 0$, the Lemma allows us to claim that the solution $\xi_l(r)$ is equally divergent in that limit, as $e^{2\beta r}/r$. The radial part of ψ: $R_l(r) = e^{-\beta r}\xi_l(r) \sim (1/r)e^{+\beta r}$ is also divergent for $r \to \infty$, thus the solutions obtained for $E < 0$ are not acceptable as square-integrable functions, for arbitrary values $\beta, \gamma \in \mathbb{R}$.

However the Lemma requires that all terms of the series (20.1) should be non zero. Yet if β and γ are not just any real numbers, but happen to meet the condition:

$$\frac{\gamma}{2\beta} = n = \text{integer} \geq l+1 ,$$

then the numerator of expression (20.1) turns into zero and the series degenerates into a polynomial. By squaring both sides of the above equation, we see the energy becomes quantized by precisely that integer number n:

$$E_n = \frac{-1}{(4\pi\varepsilon_0)^2} \frac{mZ^2e^4}{2\hbar^2} \frac{1}{n^2} \qquad n = l+1, l+2, \cdots \infty ,$$

which coincides *exactly* with the expression found in Chapter 2.3 from the quantization of the reduced classical action $S_0 = nh$. However we now have a complete understanding over the role played by angular momentum in the Hydrogen atom. Firstly, we see that the ground state corresponds to $l = 0$, having zero electron orbital angular momentum.

Secondly, we see that the list of allowed energy values: $\{E_n, n = 1, \infty\}$ is itself *independent* of l, the l values simply defining the minimum energy to be picked. As expected, the latter increases with increasing rotational energy. Once n is fixed, the ensuing list of allowed values of l is deduced:

$$l = 0, 1, \cdots n-1 .$$

Given that each value of l carries with it an energy degeneracy, due to the $2l + 1$ possible values of m_l, we derive the total degeneracy of each energy level of Hydrogen, as the sum of the first $n - 1$ odd integers:

$$g_n = \sum_{l=0}^{n-1} (2l+1) = n^2 .$$

The energy degeneracy of the different values of l, for a given n, must not be seen as a characteristic feature of radial potentials. Quite the contrary, it is an *accidental* mathematical peculiarity that only occurs in two concrete potentials, those proportional to $-1/r$ (Coulomb) and r^2 (harmonic oscillator, to be discussed in Chapter 27 [1]). Any other potential $U(r)$ would have generated energies E_{nl} dependent on *both* n and l indices.

---

[1] the law $l = 0, 1, \cdots n-1$ is not even valid for the harmonic oscillator, as we shall see.

Notwithstanding the above, the radial part of the wave function does depend on both indices n and l, whatever the potential. In the case of the Hydrogen atom it is governed by the *Laguerre polynomials* of degree q, defined by the formula:

$$L_q(x) \equiv \frac{1}{q!} e^x \left(\frac{d}{dx}\right)^q \left(e^{-x} x^q\right).$$

Table 20.1 provides the Laguerre polynomials up to $q = 5$. Obtaining the radial function $R_{nl}(r)$, of degree $n - l - 1$, requires to differentiate $p = 2l + 1$ times the Laguerre polynomial of index $q = n + l$. Which is formally expressed as:

$$L_{q-p}^p = (-1)^p \left(\frac{d}{dx}\right)^p L_q(x),$$

that includes a global sign that, in the case of Hydrogen, is always negative, $p = 2l + 1$ being an odd integer.

The polynomials are not evaluated at the radial distance in meters, but at a dimensionless distance, measured in units of the Bohr radius $\rho \equiv 2Zr/(na_0)$, given that $\beta = Z/(na_0)$. The final functions that define the states $|E_n l m_l\rangle$ for Hydrogen are:

$$\psi_{nlm_l}(r\theta\phi) = N_{nl} \, e^{-Zr/(na_0)} \, L_{n-l-1}^{2l+1}(\rho) \, \rho^l \, Y_l^{m_l}(\theta\phi),$$

with the normalization factor:

$$N_{nl} = \left[\left(\frac{2Z}{na_0}\right)^3 \frac{(n-l-1)!}{2n(n+l)!}\right]^{1/2},$$

where the previously obtained expression for the Bohr radius a_0 (with $Z = 1$, by definition) is reminded:

$$a_0 \equiv \frac{4\pi\varepsilon_0 \hbar^2}{me^2}.$$

The set of eigenfunctions $\psi_{nlm_l}(r) \equiv |nlm_l\rangle$ of the Hydrogen atom are an orthonormal basis of the Hilbert subspace of $L^2(\mathbb{R}^3)$ formed by $E < 0$ states. Which is expressed, using the scalar product in 3D (triple integral) by the following orthonormality relations:

$$\langle nlm_l | n'l'm'_l \rangle = \delta_{nn'} \, \delta_{ll'} \, \delta_{m_l m'_l}.$$

The radial functions are normalized as: $\int_0^\infty r^2 |R_{nl}|^2 dr = 1$. Note the above orthonormality relations do not necessarily mean that the radial functions alone are *always* orthogonal to each other.

In short, the wave functions of the stationary states of hydrogen (with $E < 0$) can be written as:

$$\psi_{nlm_l}(r\theta\phi) = R_{nl}(r) Y_l^{m_l}(\theta,\phi)$$

We provide in Tables 20.2 and 20.3 all the radial functions up to $n = 3$, including their normalization constants. In Figure 20.1 we show the electron probability density for the $3s$ state of Hydrogen, and the radial wave functions of all states with $n = 3$, using the polynomials and functions of the previous tables.

The eigenfunctions of hydrogen *are not valid* for multielectron atoms, even approximately. While the radial symmetry is nearly met by the field experienced by the electrons, the Coulombian degeneracy does not hold in this case. Orbitals are still described by spherical harmonics $Y_l^{m_l}$, following the standard notation $ns, np, nd, nf, ng \cdots$, for $l = 0, 1, 2 \cdots$, with *non degenerate* energies E_{nl}, where n stands for a radial excitation number.

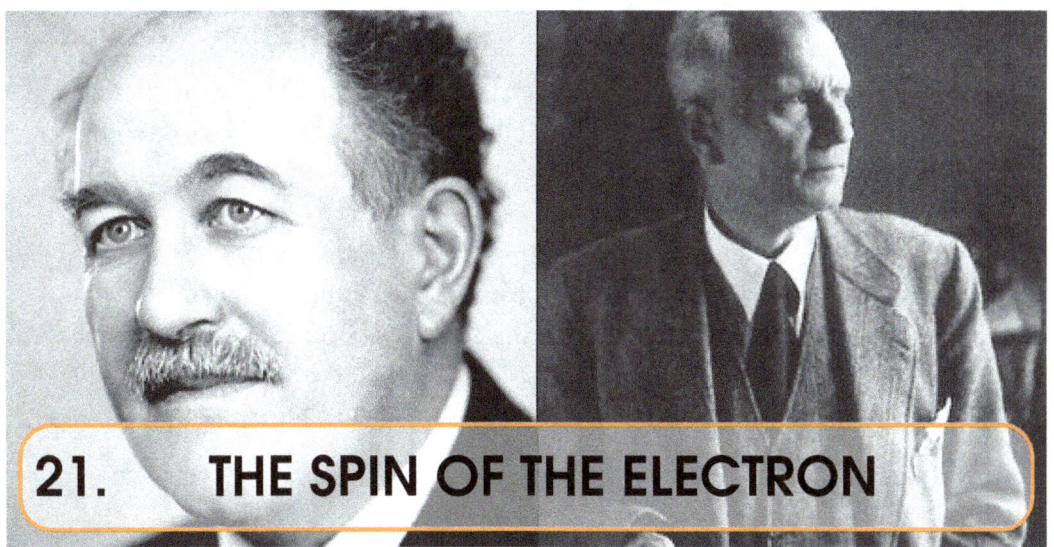

21. THE SPIN OF THE ELECTRON

We shall now describe in detail the experiment performed in 1922 by the German physicists Otto Stern and Walther Gerlach, that revealed the electron spin, a cornerstone in the consolidation of Quantum Mechanics as we know it. Its interpretation as an intrinsic angular momentum of the electron is attributed to the Dutch physicists George Uhlenbeck and Samuel Goudsmit in 1925.

What was actually measured in the above experiment was the magnetic moment of individual atoms of Silver (Ag), electrically neutral, that freely traveled in vapour form. The electronic configuration of Silver is that of the Krypton atom (a noble gas) with an additional complete d orbital (where all values of m_l with $l = 2$ are realized), plus one external $5s$ electron with $l = 0$ (zero orbital angular momentum): $[Kr]d^{10}5s^1$. Hence the magnetic moment of the Ag atom exactly equals the magnetic moment originated by the spin of that $5s$ electron. Should the magnetic moment throw a zero value, the electron would have no spin. Of course, the other electrons may also have spin, but these exactly compensate each other, within the remaining orbitals.

The mere application of a B-field on a free magnetic dipole $\boldsymbol{\mu}$ is not able to produce any resultant force on it. We assume the dipole finds itself in the quantum limit, thus the value of $|\boldsymbol{\mu}|$ cannot be altered. It simply rotates about the magnetic field B over the edge of a cone, in a phenomenon known in electromagnetism as *Larmor precession*, while its energy $-\boldsymbol{\mu}B$ remains constant.

The precession frequency specifically measures the gyromagnetic ratio $|g_e|$, being independent of the angle between these two vectors [1]. For a Bohr magneton μ_B, it would be $\omega_0/(2\pi) = \nu_0 = 13996\ MHz$ for a 1 Tesla magnetic field, if it were $|g_e| = 1$, as seen in Chapter 18.

It was beyond their reach for Stern and Gerlach in 1922 to set up a resonance experiment with microwaves, and what they did was to deviate the free dipole from its path by applying a net force, from an *inhomogeneous* magnetic field. The 3D force such a field exerts on the free dipole can be assessed as the spatial gradient of the energy (where μ is assumed to be constant):

$$\boldsymbol{F} = -\nabla U = \nabla(\boldsymbol{\mu} \boldsymbol{B}) = (\boldsymbol{\mu} \cdot \nabla)\boldsymbol{B} = \left(\mu \frac{\partial B}{\partial x}, \mu \frac{\partial B}{\partial y}, \mu \frac{\partial B}{\partial z}\right)$$

Every electromagnet needs an iron core to shape the field, and by setting up an asymmetric geometry, such as the one indicated in Figure 21.1, it is not difficult to create field lines that are *divergent* from a certain space region (see the red arrows originating from the edge, in Figure 21.1). If the atoms are incident on the magnet symmetry axis, the dominant force on the dipole only depends on the field gradient along the vertical direction (z-coordinate): $F_z = \mu_z \partial B_z / \partial z$. Note the force may be positive (upwards), or negative (downwards), depending on the product of signs of the two quantities involved.

It is easily seen that Larmor precession has the effect of averaging to zero the mean values of specific gradients $\langle \mu_{x,y} \partial B_{x,y}/\partial(x,y,z)\rangle = 0$. The gradients $\partial B_z/\partial x$ and $\partial B_z/\partial y$ may still result in a lateral displacement of the dipole, to the extent that the magnet is not exactly left/right symmetric, but it will be clearly distinguished from the main effect.

The velocity v of the Ag atoms can be calculated on average from their Maxwell distribution, as for any ideal gas: $(1/2)mv^2 = (3/2)k_BT$, with m being the atom mass, k_B Boltzmann's constant and T the absolute temperature of the oven. Under a constant force, with acceleration $a = F/m$, the atoms go through a parabolic trajectory, that produces a displacement at the magnet end: $\Delta z = (1/2)at^2 = \mu_z(\partial B/\partial z)L^2/(6k_BT)$ that is independent of the atom mass, with L being the magnet length, and $t = L/v$.

---

[1] see for example "The Magnetism of Matter", The Feynman Lectures on Physics Vol II, Ch. 34, http://www.feynmanlectures.caltech.edu/

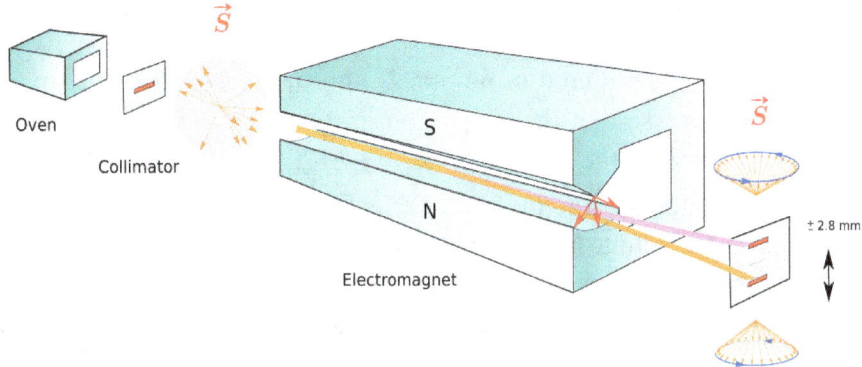

Figure 21.1: *The Stern-Gerlach experiment performed in 1922. The field gradient was $\partial B/\partial z = 10\,Tm^{-1}$, with a field $B = 1\,T$, the electromagnet length was $L = 1\,m$, and the temperature of the silver vapor $T = 400K$. The beam separation was $\pm 2.8\,mm$ on each direction. The random orientation of the incoming electron spin is also shown, as well as the spin Larmor precession on exit, once the spin has already been oriented, on each of the separated beams.*

Two key aspects deserve to be mentioned with regard to experiments of the Stern-Gerlach type: one is that they could never be performed with electrically charged particles (such as electrons), due to the Lorentz force, much larger than the above. And the other is that we are not talking about manipulating individual atoms (conceivable with today's technologies), but about the average displacement of a continuous stream of them.

Under the assumption that the Bohr magneton μ_B would define the scale of the magnetic moment, and having achieved a field gradient of $\partial B_z/\partial z = 10\,Tm^{-1}$, Stern and Gerlach expected deviations of the collimated beam of a few *mm*, as easily checked numerically from the above data.

Quantum Mechanics, as we have shown in Chapter 16, grants only two possibilities:

- no deviation of the atom beam (zero electron spin)
- an *odd* number $(2s+1)$ of spots on the screen, always showing a central one. From the observed symmetric pattern, the correct value of s would be inferred.

However, the observed results *were at variance with both assumptions*: only *two* spots were seen, deviated by $\pm 2.8\,mm$ on each direction, with no sign of a central shot, as shown in Figure 21.1.

The only way to accomodate this result within the *broader* version of Quantum Mechanics that we outlined in Chapter 18, is the following:

- the value of the quantum number s appears to be: $\mathbf{s = 1/2}$ (that is half-integer), so that the projection of the electron intrinsic angular momentum over the B-field is: $S_z = \pm\hbar/2$, i.e., $\mathbf{m_s = -1/2, +1/2}$.
- the gyromagnetic factor for the electron spin is $\mathbf{g_e = -2}$, as it is necessary to admit. Indeed, the observed deviation of 2.8 *mm* does not match numerically with $\mu_B/2$, that could naively be expected, but with a *full* Bohr magneton μ_B, as easily checked from the given data (minus sign comes from the electron charge).

Let us recall the reason why the projection of the angular momentum of a particle onto a given axis (the B-field in this case) should be an *integer* multiple of $\hbar$: the wave function ψ must be a continuous and directionally *differentiable* function in $\mathbb{R}^3$, and have no cuts. Otherwise, the probability density to find the particle at a given point, or the mean values of observables, might not be single-valued. In this case ($e^{i\phi/2}$) the directional limits for the azimuth $\phi \to 0, 2\pi$ of the wave function ψ are *not* coincidental, although the (constant) $|\psi|^2$ values are. *Two* complete space rotations (720°) need to be performed for the wave function to be continuous and *single-valued*. Which requires extending the domain of the azimuth to the interval $[0, 4\pi]$, as has been mathematically represented in Figure 21.2.

The evidence of the Stern-Gerlach experiment is incontrovertible, when it comes to accept as valid this kind of wave functions, with half-integer spin ($s = 1/2$). But we must really think at the same time about the physical meaning behind this mathematical idea: after a space rotation of 360 degrees, the quantum state of an electron does not remain unchanged, but it flips sign: $\psi \to -\psi$.

A phase change of π may give rise to destructive interference between electrons, as we shall see in the next chapter, and it is bound to have well observable physical consequences.

Both the electron spin $s = 1/2$ and the factor $g_e = -2$ were successfuly interpreted in 1932 by the British physicist Paul M. Dirac, as a consequence of his *relativistic* equation for the electron motion, the *Dirac equation*, that also describes its propagation backward in time (the positron).

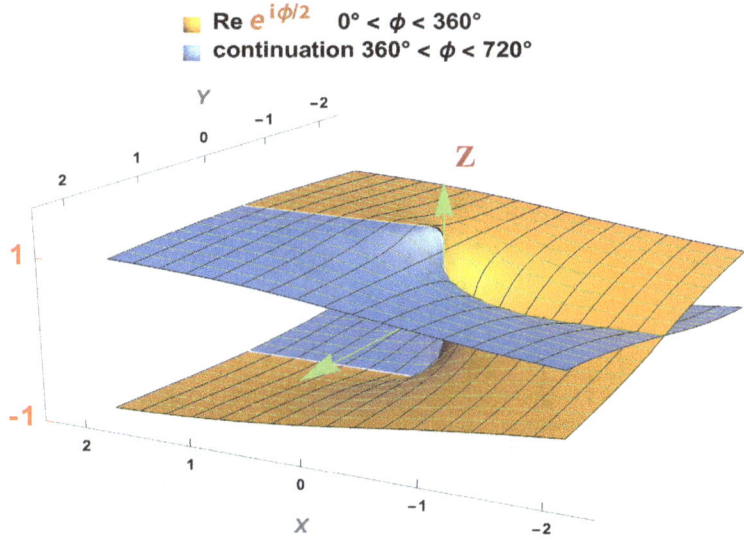

Figure 21.2: *Real part of the angular wave function $\psi = e^{i\phi/2}$ of an electron with spin oriented upwards along the Z-axis (orange surface), evaluated on the points of the horizontal plane. Note the cut (discontinuity) the function shows along the positive X-axis. That is, the function does not return to the same value after a full 360° rotation, but it flips sign: $\psi \to -\psi$. The function continuation in a second turn (sheet) has been depicted in blue, so that that the continued function is now single-valued. Such behaviour is characteristic of all fermions with spin $1/2$. The relevant distance scale for the radial part $R(r)$ of ψ is actually unknown.*

Observation of the Larmor frequency $\omega_e = |g_e|\mu_B B/\hbar$, of the electron needed to wait until 1944, by the Russian physicist Yevgeny Zavoisky, discoverer of the electron spin resonance (ESR). Proportional to the *B*-field, it takes the value $\omega_e/(2\pi B) = \nu_e/B = 28025 MHz T^{-1}$, being the basis of multiple technological applications.

In fact, already the first precision measurements showed a small, but significant, deviation of g_e with respect to -2, and today this quantity (called $g_e - 2$ in the literature), is one of the most accurate measurements in the field of physics, with 13 digits [2]. A great deal of these digits are interpreted by Quantum Electrodynamics, including vacuum polarization, and by the Standard Model of particle physics, but not all of them.

---

[2] $g_e = -2.00231930436182(52)$, https://physics.nist.gov/cgi-bin/cuu/Value?gem

Having the Stern-Gerlach experiment shown the existence of only *two* independent quantum states for the electron spin (one of them shown in Figure 21.2), the dimension of the Hilbert space in which *all* states must reside is determined. It is standard to denote the base states as:

$$|\uparrow\rangle \equiv \begin{pmatrix} 1 \\ 0 \end{pmatrix} \quad |\downarrow\rangle \equiv \begin{pmatrix} 0 \\ 1 \end{pmatrix}.$$

Thus the most general spin state of the electron is expressed, with $c_1, c_2 \in \mathbb{C}$, as:

$$|\chi\rangle = c_1|\uparrow\rangle + c_2|\downarrow\rangle \quad |c_1|^2 + |c_2|^2 = 1.$$

We may view the Stern-Gerlach experiment as a very representative example of the measurement process in Quantum Mechanics: the application of a force on the dipole triggers an observable response on the trajectory, at some point, that induces the *collapse of the wave function*, just as we described in Chapter 11. Before the measurement, the electron spin may be oriented at random, as illustrated in Figure 21.1 [3], while μ_z is not well defined. After the measurement, *only one* of the allowed eigenvalues is realized.

We see some similarity between the electron spin states $|\chi\rangle$ and the linear polarization states previously studied, living in the Hilbert space of orbital angular momentum with $l = 1$ or spin $s = 1$ (photon). Yet in the case of spin $s = 1/2$ we can perform an intuitive representation that in the most general case with $l = 1$ we cannot do. With a single parameter $0 \leq \theta \leq \pi$, we can represent both moduli: $|c_1| \equiv \cos(\theta/2)$ and $|c_2| \equiv \sin(\theta/2)$.

Which gives cause to establish a $1 - 1$ relationship between the surface points of the unit sphere, defined by the spherical coordinates (θ, ϕ), and the electron spin states defined by:

$$|\chi\rangle = \cos(\theta/2)|\uparrow\rangle + e^{i\phi}\sin(\theta/2)|\downarrow\rangle = \begin{pmatrix} \cos(\theta/2) \\ e^{i\phi}\sin(\theta/2) \end{pmatrix}. \quad (21.1)$$

Note how critical halving the θ angle is, in spherical coordinates with range $\theta \in [0, \pi]$, for the above bijection to work out. The preceding spherical surface (S^2) is called `Bloch sphere`, in the literature, and it is represented in Figure 21.3.

---

[3] or it may also have a well defined orientation, from a previous Stern-Gerlach experiment.

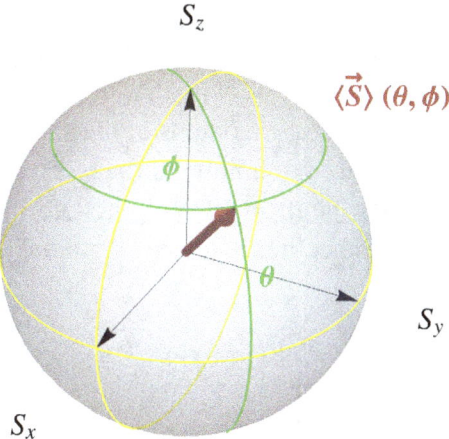

Figure 21.3: *Orientation of the mean value of the electron spin on the Bloch sphere. Each point of coordinates (θ, ϕ) on the sphere is corresponded by the quantum state determined by expression (21.1), and viceversa. All states represented here can be prepared in the lab by orienting a Stern-Gerlach magnet in the direction (θ, ϕ).*

Let us determine the form the operators must have in this two-dimensional Hilbert space (2×2 matrices), in order to represent the angular momentum $\mathbf{S} = (S_x, S_y, S_z)$. If we assume that the spin truly stands for an intrinsic instantaneous rotation, then the operators must verify the same commutation relations as those met by the infinitesimal space rotations $\mathbf{R} \equiv \mathbf{r} \times \nabla$ [4].

Indeed, the commutation relations derived from the above criterium are $[S_x, S_y] = i\hbar S_z$, nicely represented by 2×2 Hermitian matrices, with the result: $S_x = (\hbar/2)\sigma_1$, $S_y = (\hbar/2)\sigma_2$, and $S_z = (\hbar/2)\sigma_3$, as it can easily be checked:

$$\sigma_1 \equiv \begin{pmatrix} 0 & 1 \\ 1 & 0 \end{pmatrix} \quad \sigma_2 \equiv \begin{pmatrix} 0 & -i \\ i & 0 \end{pmatrix} \quad \sigma_3 \equiv \begin{pmatrix} 1 & 0 \\ 0 & -1 \end{pmatrix}.$$

These are called Pauli matrices, with spin operators $\mathbf{S} = (\hbar/2)\boldsymbol{\sigma}$, where $\boldsymbol{\sigma}$ denotes the vector formed by the 3 matrices above.

---

[4] the above expression for $\mathbf{R}$ represents, through the derivatives, infinitesimal 3D rotations about each of the coordinate axes. They meet cyclic commutation relations: $[R_x, R_y] \equiv R_x R_y - R_y R_x = -R_z$, that are therefore *geometric properties* of spatial rotations. These translate themselves to $[S_x, S_y] = i\hbar S_z$ for spin, as for any other form of angular momentum $\mathbf{r} \times \mathbf{p}$, since $\mathbf{p} = -i\hbar \nabla$ in Quantum Mechanics.

Chapter 21. THE SPIN OF THE ELECTRON

It is an interesting exercise, full of physical meaning, to show that the real vector: $\langle S \rangle_\chi = \langle \chi|S|\chi \rangle$ (mean value of the electron spin), points precisely towards the point (θ, ϕ) of the Bloch sphere that defines the state $|\chi\rangle$, and is represented in Figure 21.3. Note that the radius of the Bloch sphere: $|\langle S \rangle_\chi| = \hbar/2$ is shorter than the modulus of the spin vector: $|S| = (\sqrt{3}/2)\hbar$.

Such spin $1/2$ states are not unique to the electron, but equally happen in the proton and in the neutron. The spin state may change from one space point $\mathbf{r} \in \mathbb{R}^3$ to another, hence the most general (non relativistic) wave function is given by:

$$\chi(\mathbf{r}) = \begin{pmatrix} \psi_+(\mathbf{r}) \\ \psi_-(\mathbf{r}) \end{pmatrix} = \psi_+(\mathbf{r})|\uparrow\rangle + \psi_-(\mathbf{r})|\downarrow\rangle,$$

subject to the normalization condition:

$$\int |\chi(\mathbf{r})|^2 d^3\mathbf{r} = \int |\psi_+(\mathbf{r})|^2 d^3\mathbf{r} + \int |\psi_-(\mathbf{r})|^2 d^3\mathbf{r} = 1.$$

This means that whenever the particle reaches the point $\mathbf{r}$ through two different paths A and B, with states χ_A and χ_B (possibly having a phase difference), quantum interference will happen between them, yielding an observed particle probabiltiy density at $\mathbf{r}$:

$$|\chi_A(\mathbf{r}) + \chi_B(\mathbf{r})|^2 = |\chi_A(\mathbf{r})|^2 + |\chi_B(\mathbf{r})|^2 + 2\,\mathrm{Re}\,[\chi_A^*(\mathbf{r})\chi_B(\mathbf{r})].$$

It is interesting to know the mathematical form in which space rotations of angle α about the Z-axis act on the spin states:

$$R_{z,\alpha}(\chi) = e^{-i\alpha\sigma_3/2}\chi = \begin{pmatrix} e^{-i\alpha/2} & 0 \\ 0 & e^{i\alpha/2} \end{pmatrix} \chi(\theta,\phi) = e^{-i\alpha/2}\begin{pmatrix} \cos(\theta/2) \\ e^{i(\phi+\alpha)}\sin(\theta/2) \end{pmatrix},$$

that corroborates the expected sign flip for $\alpha = 2\pi$: $R_{z,2\pi}(\chi) = -\chi$.

Such rotations are attained, for instance, on application of a magnetic field B on the Z-axis, through the Larmor precession.

The above sign flip has been revealed in a direct manner by *neutron interference* experiments, where the observed neutron yield, as a function of the B-field, shows the necessity of a relative 4π rotation in order to have constructive interference [5].

---

[5] see the dedicated exercise proposed in the course. The original experiments were performed in 1975, H. Rauch et al. Phys. Lett. A54, 425. (1975) and S.A. Werner et al. Phys. Rev Lett. 35, 1053. (1975).

22. THE PAULI EXCLUSION PRINCIPLE

Quantum Mechanics must not only describe the motion of individual particles of mass m, but also that of *ensembles* of many identical particles. But surprisingly, it proves unable to solve, according to its own principles, a generic ambiguity that arises when it comes to define the quantum state of two particles, which drastically affects the energy of the system.

When solving Schrödinger's equation, it becomes necessary to define the quantum state ψ of *two* electrons that occupy positions r_1 and r_2 in space. Let us assume the electrons have different individual states ψ_a and ψ_b (e.g. one lying on an *s*-orbital and the other on a *p*-orbital). If we take into account that both electrons are *identical*, they will be able to occupy *simultaneously* regions of space (where their wave functions overlap), a situation which does not happen at all in Classical Mechanics. The problem is then raised how to calculate the probability density for two detectors located at r_1 and r_2 to fire simultaneously.

It is clear that the notion of the wave function ψ needs to be extended to the coordinate space $\mathbb{R}^3 \times \mathbb{R}^3 = \mathbb{R}^6$. Or equivalently, we need a propagation amplitude for two electrons to occupy the point $(r_1, r_2) \in \mathbb{R}^6$.

We are forced to admit that: $|\psi(r_1, r_2)|^2 = |\psi(r_2, r_1)|^2$, $\forall r_1, r_2 \in \mathbb{R}^3$, since both electrons are identical, and we cannot distinguish them by their spatial location. Our first option would be taking the product of both functions: $\psi(r_1, r_2) = \psi_a(r_1)\psi_b(r_2)$. However, the option of taking $\psi(r_1, r_2) = \psi_b(r_1)\psi_a(r_2)$ would do just as well, and it is *different* from the former in $\mathbb{R}^6$ space.

We may naively think this is just a formal problem, and simply take their sum: $\psi_S(r_1,r_2) = (\psi_a(r_1)\psi_b(r_2) + \psi_b(r_1)\psi_a(r_2))/\sqrt{2}$ (*symmetric* wave function). This option meets indeed the condition required at the beginning. However, it is inevitable to think that taking the *difference*: $\psi_A(r_1,r_2) = (\psi_a(r_1)\psi_b(r_2) - \psi_b(r_1)\psi_a(r_2))/\sqrt{2}$ (called *antisymmetric* wave function) also meets the condition, since $\psi_A(r_1,r_2) = -\psi_A(r_2,r_1)$ and their squares are equal.

As an example, let us consider two electrons in a given multielectron atom, in different quantum states ($a \neq b$), where the Hamiltonian contains the Coulomb repulsion energy between them. It is essential to realize that both solutions ψ_S and ψ_A produce *electrostatic energies* that are sizeably different, in eV. To convince yourself, just take the solution ψ_A and bring the two electrons at r_1 and r_2 together, $r_1 \approx r_2$. From the expression above we see that $|\psi_A(r_1,r_2)| \approx 0$. Hence the electrostatic energy of these two electrons, when they get closer, is small. By contrast, no vanishing of the wave funcion for the ψ_S solution occurs, when the electrons get closer, so their average energy is larger.

Although the above qualitative result is perfectly predictable, it is noted that the calculation of the electrostatic energy requires a 6D integral where we run over all possible relative positions. We have to average the Coulomb repulsion energy between the two electrons, $H_{12} = (e^2/4\pi\varepsilon_0)(1/r_{12})$ with $r_{12} \equiv |r_1 - r_2|$, separately on the wave functions ψ_S and ψ_A. The 6D integral requires the volume element $d\tau = d\tau_1 d\tau_2$, with $d\tau_{1,2} \equiv d^3 r_{1,2}$, and let us further simplify the notation as: $\psi(1,2) \equiv \psi(r_{1,2})$.

Thus the Coulomb interaction energy is defined as the mean value of H_{12} on either ψ_A or ψ_S, and we arrive without difficulty to the expression:

$$\int \psi^*_{S,A} H_{12} \psi_{S,A} d\tau_1 d\tau_2 = C \pm K, \qquad (22.1)$$

where C and K are the following 6D integrals [1]:

$$C \equiv \iint |\psi_a(1)|^2 d\tau_1 H_{12} |\psi_b(2)|^2 d\tau_2 \qquad K \equiv \iint \psi_a^*(1)\psi_b(1)d\tau_1 H_{12} \psi_b^*(2)\psi_a(2)d\tau_2.$$

The physical meaning of the first integral is evident: it is the electrostatic repulsion energy between the a and b orbitals. It is easy to understand the electron charge being distributed in space according to its wave function, as it occurs for a continuous charge distribution.

---

[1] it can easily be shown, upon exchange of the integration variables $1 \leftrightarrow 2$, that $K = K^*$, and also that $K > 0$.

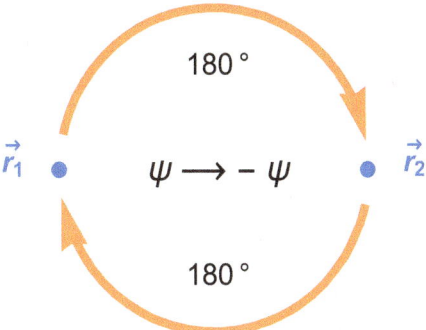

Figure 22.1: *The exchange of two identical electrons located at positions* $\mathbf{r}_1$ *and* $\mathbf{r}_2$, *through two non-intersecting paths, is equivalent to a mutual 360° rotation, that inverts the sign of the wave function.*

However there is no way to provide a classical interpretation of the K integral, because it is a *genuine quantum effect*. This integral reflects the crucial fact that *both* particles are occupying *simultaneously* each space point r.

We see it is precisely the quantity $2K$ that makes the difference between the symmetric state ψ_S and the antisymmetric state ψ_A, when it comes to calculate the Coulomb repulsion energy. It is called exchange energy in the literature, and it drives a real force on the electrons that actually governs all atomic and molecular physics. It means a *non local* force of pure quantum origin that cannot be expressed as a potential energy.

Let us keep in mind that it is indeed the Coulomb interaction that stores most of the energy released in chemical reactions. Coming back to the general question: if we have two identical particles in a given physical problem, which wave function should be used in our Hamiltonian, ψ_S or ψ_A? Surprising as it looks, the principles of Quantum Mechanics do not provide a universal answer to this question.

In the specific case of the electron, a particle of spin 1/2, we should recall the peculiar property that its wave function *flips sign* on a full space rotation of 360°: $\psi \rightarrow -\psi$. Just assume the first electron is located at point 1 and the second electron at point 2. The exchange of their positions can then be performed in two steps: transfer of the first electron from point 1 to point 2, and transfer of the second electron from point 2 to point 1, through non-intersecting paths, as indicated in Figure 22.1.

Each of these operations mean a space rotation of 180°, and the joined operation is equivalent to a *mutual* 360° rotation. Which suggests the wave function must flip sign, as discussed in the previous chapter. Thus *we must choose specifically the antisymmetric solution*, for the wave function of all pairs of identical particles of spin 1/2.

Admitting that the wave function of two electrons must always be antisymmetric has an immediate consequence, in the particular case that the quantum states a and b are identical: $a = b$. The equation follows in this case: $\psi_A(r_1, r_2) = \psi_a(r_1)\psi_a(r_2) - \psi_a(r_1)\psi_a(r_2) = 0$. Or equivalently: *in a given physical system, two electrons can never share the same quantum state*. This is the famous exclusion principle, formulated in 1925 by the Austrian physicist Wolfgang Pauli for the electron. In 1940 Pauli and other authors introduced the *spin-statistics relation*, and the antisymmetry principle, that takes up the idea that all particles of spin 1/2 must have a totally antisymmetric wave function.

Let us now consider specifically the *spin states* two electrons may have. Be reminded that spin states reside in a (dimension two) Hilbert space that is distinct from that of spatial excitations. Taking into account that both electrons are *identical*, and retracing our previous discussion on two-particle states, their combined state must be either *symmetric* or *antisymmetric* (note this consideration is independent from their spin 1/2 condition).

It is clear that when both electrons have their spins aligned in the same direction (either up or down the Z-axis), they are in a symmetric state: $|\uparrow\uparrow\rangle$ or $|\downarrow\downarrow\rangle$. However, when their spins are aligned in opposite directions, two *distinct* quantum states come up:

$$\chi_T = \frac{1}{\sqrt{2}}(|\uparrow\downarrow\rangle + |\downarrow\uparrow\rangle) \quad \chi_0 = \frac{1}{\sqrt{2}}(|\uparrow\downarrow\rangle - |\downarrow\uparrow\rangle),$$

that are called in the literature triplet and singlet states, respectively.

Note the two-particle spin states live in a $S^2 \times S^2$ Hilbert space (two Bloch spheres), endowed with a scalar product whereby states can be normalized $\langle \alpha\beta | \alpha\beta \rangle = 1$, defined as: $\langle \alpha_1\beta_1 | \alpha_2\beta_2 \rangle = \langle \alpha_1 | \alpha_2 \rangle \cdot \langle \beta_1 | \beta_2 \rangle$.

It is important to realize at this point that the antisymmetric character of the two-electron wave function we have just established, actually refers to the *totality* of their quantum state: not only the spatial part, but also the spin component. Given that the total wave function can be written as the product of both components, only two possibilities are left: either $\Psi = \psi_S \chi_A$, or else: $\Psi = \psi_A \chi_S$.

Notice that the two possibilities above no longer arise from any ambiguity of Quantum Mechanics, but from the fact that electron spins can freely orient themselves in arbitrary directions (in either χ_S or χ_A states).

It can now be understood why in the field of molecular and atomic physics we find wave functions that are either spatially symmetric ψ_S, or spatially antisymmetric ψ_A, thus showing large differences in electrostatic energy between them. What the antisymmetry principle tells us is that the ψ_S wave functions (higher electrostatic energy) are always associated with *singlet* spin states χ_0, whereas ψ_A states (lower electrostatic energy) are associated with *triplet* states χ_T, when the electron spins are opposite.

The two-electron spin states also carry energy differences, due to the dipole-dipole interaction of their magnetic moments, but these are in general much smaller than those induced by the spatial symmetry, that we have analysed in equation (22.1), as originating from the Coulomb field.

When more than two identical particles are involved in the system, the antisymmetric wave functions ψ_A we have seen can easily be extended to *determinants* formed by the individual quantum states. Independently, the antisymmetry principle also extends to the half-integer spin particles that are obtained by addition of multiple $1/2$-spins ($3/2, 5/2, \cdots$).

In the ground state of the Helium atom ($Z = 2$) we find both electrons in a $1s$ orbital, thus having a spatially symmetric wave function ψ_S, which forces the spin state to be the *singlet* χ_0. The electrostatic energy of the $1s$ orbital with itself causes the two integrals C and K in equation (22.1) to be equal, yielding the energy $2C$. The integration exercise is instructive and not too long, leading to the result: $2C = (5/2) \times 13.6\ eV = +34\ eV$. Adding the above to the naive sum of ground state energies with a He nucleus, yields $-2Z^2 \times 13.6\ eV + 34\ eV = -108.8\ eV + 34\ eV = -74.8\ eV$, which is in reasonable agreement with the measured He ground state energy, namely: $-79.0\ eV$.

Another remarkable example of the antisymmetry principle comes with the so-called *Hund rule* of atomic and molecular physics, by which all electrons occupying a half-filled orbital tend to align their spins in the *same* direction $|\uparrow\uparrow\uparrow \cdots\rangle$. In this way a *lower electrostatic energy* is attained, due to the spatially-antisymmetric wave function ψ_A (e.g. of p_x, p_y, p_z orbitals), as previously explained (the spin wave function being obviously symmetric χ_S).

The implication of the antisymmetry principle goes far beyond molecules and atoms, and truly affects the whole of physics. To be aware of that, let us first know which particles have half-integer spin, and which have integer spin. The former are called `fermions` and the latter are called `bosons` in the literature [2].

Fermions are the particles that we usually call *matter*, the most relevant of them being:

- the electron
- the quarks
- the proton and the neutron, formed by quarks and gluons

Bosons are the particles that we usually call *radiation*, plus other particles composed of fermions that happen to have integer spin. Some of the most relevant examples are:

- the photon
- the phonon
- the gluon
- the Helium atom

The statistical ensembles of many identical particles are called *fermionic systems*, or *bosonic systems*, in the literature. When their interaction can be neglected (other than for symmetry reasons), they are often called fermion or boson *gases*. Bosons do not have the peculiar property of their wave function flipping sign after a 360° rotation, and therefore *pairs of identical bosons must have a totally symmetric wave function*.

The very fact of not being compelled to obey the exclusion principle renders boson systems physical properties that are *radically different* from those of fermion systems. Most characteristically, nothing prevents many particles lying in the ground state of the system. It is well known that such configuration would never happen with the electrons of an atom, but it does happen in a `laser`, formed by many photons sharing equal value of their momentum p (totally monochromatic) and of their spin ($S_z = \pm\hbar$).

---

[2] in recognition to the Italian physicist Enrico Fermi (1901-1954) and to the Indian physicist Satyendra Nath Bose (1894-1974), respectively.

In a similar way we have, at low temperatures, the phenomenon of superfluid Helium, and that of superconductivity, both originating from the properties of specific bosonic systems. Also the photon gas that fills the cavity of an oven in thermal equilibrium with its walls at temperature T, is a boson system. The photon energy distribution is governed by *Planck's law*, slightly differing from Maxwell's energy distribution for an ideal gas, due to the bosonic character of the photon. A particular case of the photon gas is found in the cosmic microwave background (CMB) that fills the whole of the universe at 2.7 K.

As fermionic systems of interest, besides atoms and molecules, we shall mention the atomic nuclei, the electron gas that characterizes all metals, the neutron gas that governs the shielding calculations of nuclear reactors, and the neutron stars in astrophysics. A great diversity of fermionic, and bosonic, systems can be found in physics.

23. THE PARTICLE IN A BOX

Let us examine a case of Schrödinger's equation where the analytical simplicity does not diminish its extensive use and profound significance. We are referring to a particle enclosed between 6 impenetrable parallel walls, located at a common distance L.

We begin by solving the problem in 1D. The particle moves between the points: $x \in (0, L)$ with $U(x) = 0$, and we assume the force $F = -\partial U / \partial x$ is infinite (and inward directed) at both ends: $U(0) = U(L) = \infty$. Eigenvalue equation $H\psi = E\psi$ is assessed: $-\hbar^2 \partial^2 \psi / \partial^2 x = 2mE\psi(x)$ and the general solution is attained:

$$\psi(x) = Ae^{-ikx} + Be^{ikx} \quad A, B \in \mathbb{C},$$

with $k = \sqrt{2mE}/\hbar > 0$. The physical meaning of each of the above terms is clear: the particle travels *simultaneously* toward the right and toward the left, according to the Feynman propagator. The boundary conditions are:

$$\psi(0) = A + B = 0 \qquad \psi(L) = Ae^{-ikL} + Be^{ikL} = 0,$$

which mean $A(e^{-ikL} - e^{ikL}) = A[-2i\sin(kL)] = 0$, the solution being $k_n L = n\pi$ with $n = 1, 2, \cdots \infty$. Hence the normalized eigenfunctions for the stationary states are:

$$\psi_n(x) = C_n \sin(k_n x) \qquad C_n = \sqrt{\frac{2}{L}},$$

where C_n are the normalization constants (that happen to be independent of n), that garantee: $\int_0^L |\psi_n|^2 dx = |C_n|^2 \int_0^L \sin^2(n\pi x/L) dx = 1 \; \forall n$.

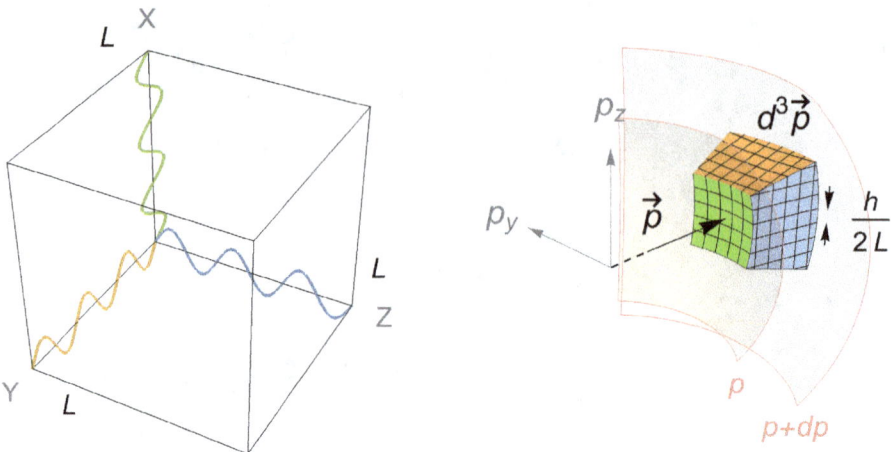

Figure 23.1: *a) One-dimensional wave functions projected onto each coordinate axis, for a particle enclosed in a cube of side L. b) Momentum vector in 3D space of the particle enclosed in the cube on the left, that points to the center of elementary cells of side $h/(2L)$. The volume element in momentum space is also shown, in spherical coordinates, containing a finite (and very large) number of elementary cells.*

Thus the energies of the stationary states are:

$$E_n = \frac{\pi^2 \hbar^2}{2mL^2} n^2 \qquad n = 1, 2, \cdots \infty.$$

The energy of the ground state E_1 could have been estimated by using the uncertainty principle $\Delta x \Delta p \sim \hbar$, and grows infinitely when the two walls get very close ($L \to 0$), as it was to be expected.

We see that the energy levels are proportional to the squares of the natural numbers n^2, and split up from each other as n increases. However, their separation *relative to the baseline energy*, becomes infinitely small as $n \to \infty$:

$$\frac{E_{n+1} - E_n}{E_n} = \frac{(n+1)^2 - n^2}{n^2} = \frac{2n+1}{n^2} \to 0,$$

which should also be expected in the classical limit $n \to \infty$, similarly to what we saw for the Hydrogen atom (despite the opposite behaviour with n^2). Note the energy quantization with n^2 is actually a consequence of the quantization of the *modulus* of the momentum, as integer multiples of a fixed quantity: $|p_n| = n\pi\hbar/L$.

Another way to express the above is saying that the length of the cavity must be an *integer multiple of the De Broglie half-wavelength* of the particle: $L = n \cdot (\lambda_n/2)$, as shown in Figure 23.1.

As to the wave functions, they are actually identical to those found in the problem of stationary waves within the same cavity, since for $U(x) = 0$ the wave equation and the Schrödinger equation show the same solutions for standing waves [1].

It is remarkable that in the ground state, the particle meets its maximum probability density at the center of the cavity $x = L/2$. This is opposite to what happens in Classical Mechanics, where random pictures of the position would produce a uniform distribution.

The extension of the above result to 3D is straightforward, in cartesian coordinates. The boundary condition is now that $\psi(r)$ must be zero at the 6 walls of the cavity. It is obvious that we can build such solutions for the stationary states as:

$$\psi_{n_1 n_2 n_3}(r) = \left(\frac{2}{L}\right)^{3/2} \sin\left(\frac{n_1 \pi x}{L}\right) \sin\left(\frac{n_2 \pi y}{L}\right) \sin\left(\frac{n_3 \pi z}{L}\right),$$

in correspondance with the 3D energies:

$$E_{n_1 n_2 n_3} = \frac{\pi^2 \hbar^2}{2mL^2} \left(n_1^2 + n_2^2 + n_3^2\right) \quad n_{1,2,3} = 1, 2, \cdots \infty.$$

Note that now the energy of the ground state E_{111} is three times larger than in 1D. This could be forseen, since confinement in each of the 3 space dimensions generates momenta that add up quadratically, as arising from the uncertainty principle.

It is important to emphasize that none of the 3 integers is allowed to be zero, since a wave function $\psi(r) = 0$ would not be correctly normalized. We also see that in 3D each energy level acquires a quantum degeneracy it did not have in 1D, because of the different lists (n_1, n_2, n_3) of 3 integers that are compatible with a fixed value of the sum $n^2 = n_1^2 + n_2^2 + n_3^2$. The degeneracy increases with n, and it is an interesting exercise to calculate it for the lowest integers.

---

[1] note the momentum quantization law $|p_n| = n\pi\hbar/L$ is totally relativistic, hence the solutions obtained are also valid for *a photon* (mass $m = 0$) enclosed in the box, with relativistic energy: $E_n = \sqrt{n_1^2 + n_2^2 + n_3^2}\, \pi\hbar c/L$. This is interesting, because the standing electromagnetic wave in the cavity actually represents the photon's wave function.

But the most interesting calculation, for its application to statistical mechanics and to the scattering theory of elementary particles, is that of the quantum degeneracy *in the classical limit*, when $n_i \gg 1$ for all excitation numbers. Its rationale is illustrated in Figure 23.1, where the quantum values of the momentum vector p actually point to the nodes of a 3D grid of *elementary* cubes with side $h/(2L)$, as a consequence of the momentum quantization on each direction. Even if infinitesimally small from the standpoint of integral calculus, the volume element d^3p still contains a *finite*, and very large, number of quantum states. This *integer* number is easily calculated as the ratio:

$$d\mathcal{N} = \frac{d^3p}{\left(2 \cdot \frac{h}{2L}\right)^3} = \frac{d^3p/8}{\left(\frac{h}{2L}\right)^3}.$$

The factor 2 in the denominator accounts for the two possible signs of the momentum on each coordinate axis ($\pm p_{x,y,z}$). Recall that while it is $|p|$ that is quantized, the particle may take any ($\pm$) direction, upon integration over the 3D phase-space in $\mathbb{R}^3$.

Given the volume of the cavity $V = L^3$, the above formula provides the *exact* number of quantum states per unit volume in d^3p, in the *classical limit* $n \to \infty$:

$$\frac{d\mathcal{N}}{V} = \frac{d^3p}{h^3}.$$

Note this result is 100% relativistic, since it is ultimately based on the De Broglie *wavelength* of the particle inside the box (its frequency not being used). Which means it can be used for a gas of ultrarelativistic or semirelativistic particles (photons, positrons, neutrinos, etc), filling the volume V. It can also be used for a *free electron gas* in a conductor.

Let us now particularize the previous expression by choosing a specific energy/momentum relationship, namely the non relativistic $E = p^2/(2m)$, suitable for a molecular gas at room temperature [2]. Then $dE = (p/m)dp$, and $d\mathcal{N}$ gives the number of quantum states available to each molecule in the kinetic energy interval $(E, E+dE)$.

---

[2] for low enough temperatures, the approximation we are using ($n_i \gg 1$) necessarily breaks down, since the excitation levels n_i will get close to the ground state $n_i = 1$. When this happens, the statistical analysis requires taking into account the fermionic or bosonic character of the particles filling the cavity.

We now explain the calculation of $d\mathcal{N}$ in 3D, for an interval $(p, p+dp)$ of the *modulus* of the momentum. Given the value of E, $p = |p|$ means the radius of the spherical surface $p^2 = p_x^2 + p_y^2 + p_z^2 = 2mE$, for which the spherical ring, between radii p and $p + dp$, has been represented in Figure 23.1. The volume element in spherical coordinates $d^3p = p^2 dp d\Omega$ can now be integrated over the solid angle, yielding $\int d\Omega = 4\pi$, to get the number of quantum states enclosed in the spherical ring:

$$d\mathcal{N} = \frac{4\pi V}{h^3} p^2 dp \,. \tag{23.1}$$

Which, upon substitution of $dE = (p/m)dp$, leads to our final (non relativistic) result for the total number of different quantum states that the particle, with kinetic energy in the interval $(E, E+dE)$, *may have*:

$$d\mathcal{N} = g(E)dE = \frac{4\pi V}{h^3}\left(2m^3\right)^{1/2} E^{1/2} dE \,.$$

The standard notation $g(E)dE$ has been used to signify the *quantum degeneracy* that corresponds to the *continuous* energy level $(E, E+dE)$. Note that $d\mathcal{N}$ is actually an *integer* number [3], as shown in Figure 23.1. It is the number of states that the molecule may occupy, proportional to the energy width dE, with $g(E) = d\mathcal{N}/dE$ being the *density of states per unit energy interval*. This is the contribution of each particle to the *entropy* of the system, when we have many particles. It is revealing that this quantity would be *infinite* in Classical Mechanics.

The entropy of a gas of N molecules enclosed in volume V is simply given by the natural logarithm of the total number of quantum states W these molecules may have in the volume: $S = k_B \ln W$. The Boltzmann constant k_B just grants the proper dimension of entropy in JK^{-1}. There is no doubt how to proceed for the calculation of W: the degeneracy of each energy level i must be *multiplied* for all levels, raised to the number of molecules n_i found in that particular level: $W = \prod_i (g_i^{n_i})$.

However, owing to the fact that all molecules n_i are *identical*, we cannot simply count their different permutations as different states, hence the above formula needs to be corrected in the form: $W = \prod_i (g_i^{n_i}/n_i!)$. In thermodynamic equilibrium at temperature T, the population $n_i(T)$ follows a Maxwell-Boltzmann distribution: $n_i(T) = (N/Q)g_i e^{-E_i/k_B T}$, with $Q \equiv \sum_i g_i e^{-E_i/k_B T}$.

---

[3] it strictly corresponds to the integer part of expression (23.1).

Indeed, the Maxwell-Boltzmann distribution can be *derived* generically from the criterion of *maximizing* the above entropy.

In order for the previous expression of W to be accurate, the temperature T must not be too low, the product $k_B T$ being required to largely exceed the ground state energy of the cavity ($k_B T \gg E_{111}$). Should this not be the case, the bosonic or fermionic character of the particles in the cavity needs to be taken into account specifically, for a precise determination of their entropy. This is understood because, at low temperature, De Broglie's particle wavelength becomes larger. Thus individual particle wave packets overlap, and their spatial symmetry comes naturally into play.

It is truly amazing that the Austrian physicist Ludwig Boltzmann could establish formula $S = k_B \ln W$ around 1873, yet without knowing about the Planck constant, which certainly places him among the most outstanding physicists of all time. Let us recall that entropy is strictly infinite in Classical Mechanics. Planck first (1900), and Einstein later (1905), used Boltzmann's idea of entropy to discover the h constant in the atomic oscillators, and in the electromagnetic radiation (photons), respectively.

24. THE PIECE-WISE POTENTIALS

Let us consider the case of a particle of mass m that goes through a sudden increase of potential energy, that jumps from a baseline constant $U = 0$ for $x < 0$ to a positive and constant value E_0 for $x \geq 0$, as illustrated in Figure 24.1. This simple mathematical description as a 1D step function is often used, despite the fact that the force $F = -\partial U/\partial x$ must always be *finite* and stretch over a finite space region, of length $d \neq 0$. A practical realisation is provided by an electron incident on the symmetry axis of two cylindrical capacitors, with a potential difference V, just as indicated in Figure 24.1.

Having $d = 0$ is unfeasible, owing to the necessary insulation between the two capacitors, as shown in the figure drawing. Still we are going to solve Schrödinger's equation for the stationary states assuming $d = 0$, due to its mathematical simplicity, even though such assumption may give rise to some comprehension difficulties with the solutions obtained, as will be seen.

In all problems characterized by piece-wise potentials (of which this is the simplest example) we meet plane waves $Ae^{\pm ikx}$ as solutions in the regions of constant potential (as discussed in the previous chapter), that may extend to infinity. Given that these wave functions are *ideal*, i.e. non square-integrable, it becomes necessary to provide a physical interpretation of the complex coefficient A. Its modulus can be associated with the *flux* $\phi \equiv v|A|^2 = (\hbar k/m)|A|^2$ of *particles per unit time* that flow through the point x with velocity v, in the direction indicated by the sign of the exponent.

Note that the dimension s^{-1} of ϕ is derived from the dimension m^{-1} of $|\psi|^2$, by virtue of its probabilistic interpretation in 1D. The *phase* of A would not make sense if the wave were isolated, but it does make sense when it is part of a linear combination with other waves. Then it just encodes the phase *shift* that these waves have acquired with respect to a common reference (the incident wave).

We analyse first the case $E = \hbar^2 k^2/2m > E_0$, where the incident particle has kinetic energy larger than E_0. In Classical Mechanics, it would always cross to the other side. To study what happens in reality, the simplest way is to find the stationary states as solutions to the Schrödinger equation $H\psi = E\psi$, and consider the condition that the particle is *incident from the left-hand side* as a boundary condition. We could also parametrize an incident *pulse* with a given analytic shape, and study its time evolution with the equation $i\hbar\,\partial\psi/\partial t = H\psi$. The results obtained in this case would be more realistic, since the particle could be followed in time with a certain resolution $\Delta t \neq \infty$, as it really happens in the laboratory, but on the other hand, they would depend on the pulse shape chosen.

In either case, we are interested in the probability R for the particle to bounce back, when its energy E is larger than the step height E_0, in the limit where E is known with infinite precision ($\Delta E \to 0$).

Denoting by $\psi_{1,2}$ the solutions to $H\psi = E\psi$ on the left-hand side ($x < 0$) and on the right-hand side ($x > 0$), respectively, the general solution is written as:

$$\psi_1 = Ae^{ik_1x} + Be^{-ik_1x} \qquad \psi_2 = Ce^{ik_2x} \qquad A,B,C \in \mathbb{C},$$

with $k_1 \equiv \sqrt{2mE}/\hbar$ and $k_2 \equiv \sqrt{2m(E-E_0)}/\hbar$. It is easily understood that, for consistency with the initial condition we have assumed (that the particle is incident from the left-hand side), a solution that includes a wave $\psi_2 = De^{-ik_2x}$, that propagates from right to left, standing on the right-hand side, *cannot be admitted*.

In this approach, the coefficient A should be considered as a *datum* of the problem, representing the *incoming flux* of particles, the complex numbers B and C being univocally determined from A. Since the overall phase of ψ is meaningless, we can take A as a real positive number, without loss of generality.

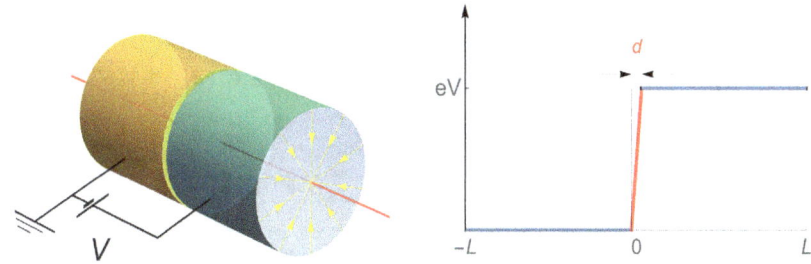

Figure 24.1: *Realistic example of a step potential created by two consecutive cylindrical capacitors, the first grounded to earth and the second set to a relative potential V. The electron is incident on the symmetry axis, where the total electric field (yellow arrows) is zero. Due to the necessary insulation between the two capacitors, it is inevitable that the potential must be biased over a certain distance $d \neq 0$.*

The reflection probability R, according to our previous discussion about the flux, is given by:

$$R = \frac{|B|^2}{|A|^2}.$$

The method to calculate B and C is simple: both ψ and $\partial\psi/\partial x$ must be *continuous* functions in $\mathbb{R}$, in particular at $x = 0$:

$$\psi_1(0) = \psi_2(0) \qquad \frac{\partial \psi_1}{\partial x}(0) = \frac{\partial \psi_2}{\partial x}(0),$$

hence:

$$A + B = C \qquad ik_1 A - ik_1 B = ik_2 C, \tag{24.1}$$

from which we deduce:

$$B = \left(\frac{k_1 - k_2}{k_1 + k_2}\right) A \qquad C = \left(\frac{2k_1}{k_1 + k_2}\right) A,$$

that lead us to the reflection probability:

$$R = \left(\frac{k_1 - k_2}{k_1 + k_2}\right)^2 = \left(\frac{1 - \sqrt{1 - (E_0/E)}}{1 + \sqrt{1 - (E_0/E)}}\right)^2. \tag{24.2}$$

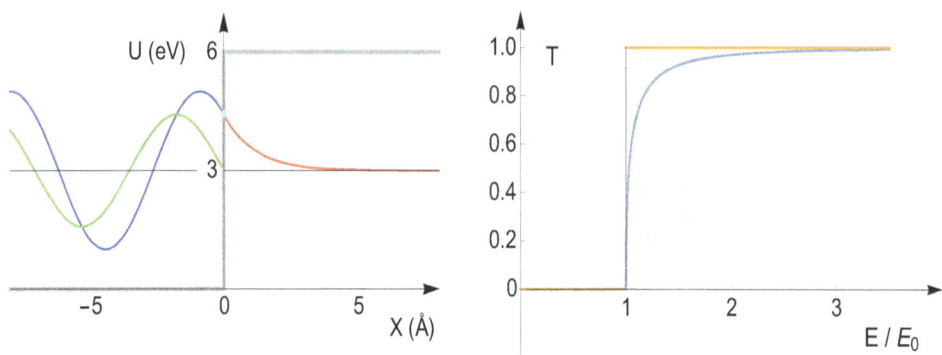

Figure 24.2: *a) Real wave function ψ_1 (blue) and ψ_2 (red) of an electron with kinetic energy $E = 3\,eV$ incident on a potential step of height $E_0 = 6\,eV$. The wave function acquires a phase shift of $\phi = -45°$ with respect to that incident on a potential step of infinite height $E_0 \to \infty$ (green), according to equations* (24.4). *The length scale is shown in Å. The wave function is not measured in energy units, so the vertical scale is shown only for reference. b) Behaviour of the transmission probability $T = 1 - R$ as function of the energy ratio (blue line) for the step potential shown on the left, according to expression* (24.2). *The prediction of Classical Mechanics has also been drawn (orange line).*

We could also have defined a *transmission coefficient* T (probability that the particle crosses to the other side), taking into account in this case the velocity decrease ($k_2 < k_1$) to evaluate the flux. As expected, both probabilities add up to one:

$$T = \frac{k_2}{k_1} \frac{|C|^2}{|A|^2} = 1 - R\,.$$

Expression (24.2) has been represented in Figure 24.2(b), as a function of E/E_0. Note this formula is symmetric upon the exchange $k_1 \leftrightarrow k_2$. This is telling us that should we have stated the initial condition as a particle incident *from the right-hand side*, the same result would have been obtained (as it can easily be checked). However, both situations are different in physical terms: in one case the particle is decelerated by making work against the electric field, whereas in the other case the field accelerates the particle to increase its kinetic energy. The probability to bounce back, in either case, is the same, as we have shown. In fact, one process is related to the *time reversal* of the other. Note the limits $R \sim (E_0/4E)^2 \to 0$ for $E \to \infty$ and $R \to 1$ for $E \to E_0$.

Even if the electromagnetic case provides the most glaring example of the step potential, there are other relevant cases. For instance, a potential drop may be originated by nuclear energy (with $E_0 < 0$ in this case). The particle could be a neutron incident on a large size nucleus, that exerts an attractive force on it (inwards) only at the nucleus surface. In this case the wave character of the neutron causes its bouncing back with a probability calculable from expression (24.2), if we know the potential drop $(-E_0)$ in MeV for that particular nucleus.

The Italian-American physicist Enrico Fermi and his collaborators made use of the above probability calculation to control the first self-sustained nuclear reaction chain, in the Chicago Pile-1 reactor, in December 1942.

Let us now discuss a paradoxical situation that arises when the above calculation is taken to the classical limit. The ratio (24.2) can be expressed as $R = [(v_1 - v_2)/(v_1 + v_2)]^2$, since it only depends on the velocity *variation*, being independent of $\hbar$. Therefore, it should also be valid in the limit $\hbar \to 0$. And this is *absurd*, for the non zero probability is undoubtedly associated with the wave character of the particle, whose De Broglie wavelength λ is proportional to $\hbar$.

The explánation of this paradox is to be found in the *ideal* assumption of an infinite force at $x = 0$, which has oversimplified the problem. Had we assumed a *finite* distance $d \neq 0$, by means of a particular line shape for the potential bias (linear or not), as indicated in Figure 24.1, and solved the equation $H\psi = E\psi$ for the stationary states, the result for R would have depended explicitely on $\hbar$. In this case, the correct result goes to zero in the classical limit: $R \to 0$ for $\hbar \to 0$, as it should [1].

We may still use formula (24.2) in Quantum Mechanics, for the step potential, subject to admitting that $d \neq 0$ always, and restricting its validity to the quantum limit, which means that the De Broglie wavelength λ of the incident particle must be *equal or greater than the distance d*:

$$\lambda \gtrsim d.$$

Obviously, it is very difficult that macroscopic bodies with extremely short λ can meet the above condition, since that would mean an unlikely precision when it comes to establish the distance d in the laboratory.

---

[1] both calculations can be found, for two specific line shapes of the potential bias, in the book "Quantum Mechanics" by L. D. Landau and E. M. Liftshitz, Butterworth-Heinemann (1977).

Chapter 24. THE PIECE-WISE POTENTIALS

Let us now consider the case of the particle energy being less than the step height, $E < E_0$. The general solution of $H\psi = E\psi$ only differs from the previous calculation by the presence of a real, decreasing, exponential function on the right-hand side:

$$\psi_1 = Ae^{ikx} + Be^{-ikx} \qquad \psi_2 = Ce^{-\alpha x} \qquad A, B, C \in \mathbb{C}, \qquad (24.3)$$

Now what we have in the exponent is $\alpha \equiv \sqrt{2m(E_0 - E)}/\hbar > 0$, and in this case we call $k \equiv \sqrt{2mE}/\hbar$. On application of the continuity conditions (24.1) we easily obtain the relations:

$$B = \left(\frac{ik+\alpha}{ik-\alpha}\right) A \qquad C = \left(\frac{2ik}{ik-\alpha}\right) A.$$

Notice that now the complex numbers B and C have acquired a phase shift with respect to A, and that the particle may now reside on the right-hand side, which is classically forbidden. However, a detailed calculation of the reflection probability R reveals it is always 100%, as classically expected:

$$R = \frac{|B|^2}{|A|^2} = \left|\frac{ik+\alpha}{ik-\alpha}\right|^2 = \frac{k^2+\alpha^2}{k^2+\alpha^2} = 1.$$

The phase shift that is produced in the reflected wave is some sort of mathematical reminder of the time the particle actually spends in the forbidden region, when $\Delta t \neq \infty$ (for non stationary cases). We have plotted in Figure 24.2(a) the superposition of the incident wave and the reflected wave, calculated as follows:

$$\psi_1(x) = Ae^{ikx} + \left(\frac{ik+\alpha}{ik-\alpha}\right) Ae^{-ikx} = \frac{2iA}{(ik-\alpha)} (k\cos(kx) - \alpha\sin(kx)).$$

The superposed function upstream is then *real*, since the overall phase is irrelevant. In fact, it can be expressed as a sine function, phase shifted by an angle ϕ that is hence uniquely determined [2]:

$$\psi_1(x) = F \cdot \sin(kx+\phi) \qquad \text{Tan}\,\phi = \frac{-k}{\alpha} \qquad F = \frac{2iA}{(ik-\alpha)} \frac{-1}{\sqrt{k^2+\alpha^2}} \qquad (24.4)$$

---
[2] recall $\sin(\alpha+\beta) = \sin\alpha\cos\beta + \cos\alpha\sin\beta$.

25. THE POTENTIAL BARRIER

Let us now proceed to the next level of piece-wise potentials: after the initial step, the potential *returns* to its baseline value $U = 0$ at a certain distance a. A real game changer, causing the quantum properties of the system to be observable in a most striking way. This case is generally known as *potential barrier* of width a. That is, $U(x) = 0$ for $x < 0$, $U(x) = E_0$ for $0 < x < a$ and $U(x) = 0$ for $x > a$.

Similarly to the single step case, we have two distinct possibilities: the kinetic energy E is above the barrier height $E > E_0$, or below $E < E_0$. Notice that throughout the following calculations, the value of E_0 may be positive ($E_0 > 0$) or negative ($E_0 < 0$), but the sign information will not be used, except for the graphical representation where we will choose the positive barrier.

From the mathematical standpoint, both cases are similar. The general solution in any region where $E < E_0$, contains real exponentials $e^{\pm \alpha x}$, contrary to complex exponentials for $E > E_0$ $e^{\pm ik_2 x}$, with corresponding values of $\alpha \equiv \sqrt{2m(E_0 - E)}/\hbar > 0$ and $k_2 \equiv \sqrt{2m(E - E_0)}/\hbar > 0$. It is clear that $\alpha = ik_2$, and given that we are dealing with complex numbers for the coefficients, we may immediately obtain the solution for $E < E_0$ from the one with $E > E_0$, by performing the above substitution in all the expressions. However, the physical meaning of the solutions is *totally* different, giving rise to entirely different phenomena, as we shall see.

Chapter 25. THE POTENTIAL BARRIER

The mathematical structure we are going to handle is general, and may easily be extended to an arbitrary number of chained potential barriers at succesive locations. Since in each piece the potential energy is constant, the general solution to the Schrödinger equation contains two terms with complex coefficients (as seen earlier with the single step). Each new barrier introduces two new constraints, namely the continuity of ψ and of $\partial \psi / \partial x$ at the new boundary point. Thus we always meet a *linear* system of complex numbers (the coefficients), that shows a unique solution. Technically, it is enough to invert a $N \times N$ complex matrix in order to define the wave function (modulus and phase) at every space point $x \in \mathbb{R}$.

Let us take as reference the case $E > E_0$, and divide the space $(-\infty, +\infty)$ in 3 pieces: $x < 0$, $0 < x < a$, and $x > a$ with wave functions $\psi_{1,2,3}$. The general solution can be written as:

$$\psi_1 = A_0 e^{ik_1 x} + A e^{-ik_1 x} \qquad \psi_2 = B e^{ik_2 x} + C e^{-ik_2 x} \qquad \psi_3 = D e^{ik_1 x}, \quad (25.1)$$

where $A_0 \in \mathbb{R}$ is the *datum* that corresponds to the flux incident *from the left-hand side* (boundary) condition. The wave numbers are: $k_1 = \sqrt{2mE}/\hbar$ and $k_2 = \sqrt{2m(E-E_0)}/\hbar$. The linear system of the *complex* coefficients A, B, C, D is determined by the 4 continuity equations:

$$\psi_1(0) = \psi_2(0) \quad \frac{\partial \psi_1}{\partial x}(0) = \frac{\partial \psi_2}{\partial x}(0) \quad \psi_2(a) = \psi_3(a) \quad \frac{\partial \psi_2}{\partial x}(a) = \frac{\partial \psi_3}{\partial x}(a) \quad (25.2)$$

leading to the expressions:

$$A_0 + A = B + C \qquad ik_1 A_0 - ik_1 A = ik_2 B - ik_2 C$$

$$B e^{ik_2 a} + C e^{-ik_2 a} = D e^{ik_1 a} \qquad ik_2 B e^{ik_2 a} - ik_2 C e^{-ik_2 a} = ik_1 D e^{ik_1 a}.$$

$$(25.3)$$

The above system of equations has a unique solution, that has been depicted in Figure (25.1) for a particular value of the energy E, the height E_0, the width a, and the mass m. Without formally proceeding to the inversion of its 4×4 complex matrix, let us focus on the D number, in order to determine the *transmission coefficient*, or probability for the particle to cross to the other side of the barrier. According to our earlier definition of the flux, this probability is given by the expression:

$$T = \frac{|D|^2}{|A_0|^2} = \frac{DD^*}{|A_0|^2}. \quad (25.4)$$

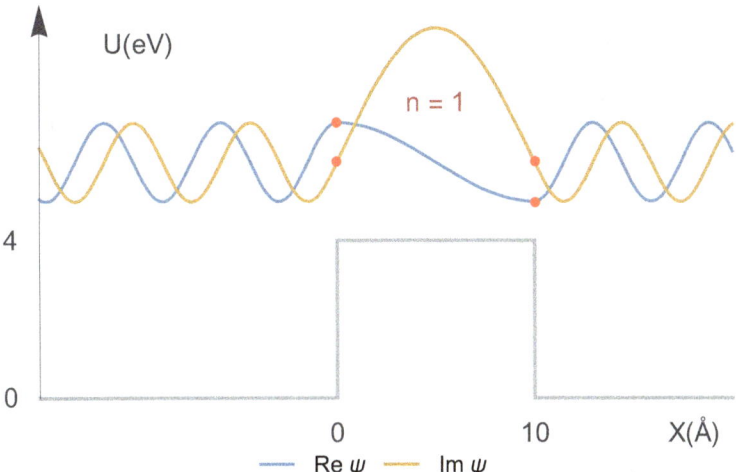

Figure 25.1: *Real and imaginary part of the wave function of an electron incident with energy $E = 4.3761\,eV$ (resonant value corresponding to $n = 1$) on a potential barrier of height $E_0 = 4\,eV$ and width $a = 10\,\text{Å}$. Note the four points (shown in red) where the continuity condition for ψ and $\partial\psi/\partial x$ stated in* (25.2) *has been applied.*

Note that the velocity in the last piece is exactly the same as the incident velocity, as shown by the equal values of k_1 in the general solution (25.1), which expresses the energy conservation.

To obtain D, it is enough to combine the two equations in the first row in (25.3) into only one:

$$2k_1 A_0 = (k_1 + k_2)B - (k_2 - k_1)C , \qquad (25.5)$$

and express the second pair of equations in the form:

$$B = \frac{k_1 + k_2}{2k_2} D e^{ik_1 a} e^{-ik_2 a} \qquad C = \frac{k_2 - k_1}{2k_2} D e^{ik_1 a} e^{ik_2 a} . \qquad (25.6)$$

Upon substitution of B and C in equation (25.5) we obtain a unique determination of the complex number D:

$$4k_1 k_2 A_0 = \left((k_1 + k_2)^2 e^{-ik_2 a} - (k_2 - k_1)^2 e^{ik_2 a}\right) D e^{ik_1 a} .$$

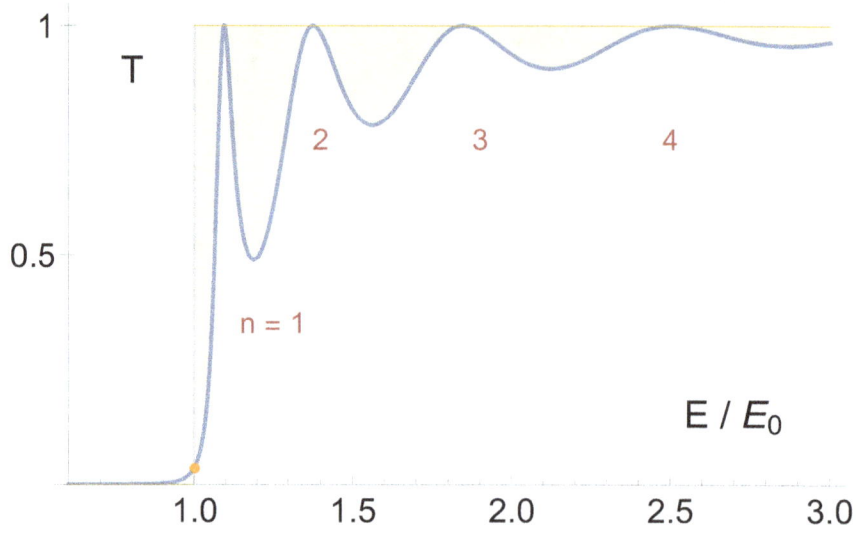

Figure 25.2: *Transmission probability for a potential barrier as function of the ratio E/E_0, according to formula (25.7). The data actually correspond to the barrier shown in Figure 25.1. The lowest resonances for $n = 1, 2, 3$ and $n = 4$ are clearly visible. Note the decreasing sharpness of the resonance with increasing n. The difference with respect to Classical Mechanics is filled with yellow color.*

Now multiplying the above expression by its complex conjugate, the difference: $(k_1^2 - k_2^2)(e^{2ik_2a} + e^{-2ik_2a})$ is isolated, that is proportional to the real function: $\cos(2k_2a) = 1 - 2\sin^2(k_2a)$. From DD^* we obtain the transmission coefficient, that is expressed according to (25.4) as a sine function:

$$T = \frac{1}{1 + \dfrac{\sin^2(k_2a)}{4\dfrac{E}{E_0}\left(\dfrac{E}{E_0} - 1\right)}} .\qquad(25.7)$$

The above expression has a physical meaning that is far-reaching. In Classical Mechanics, the particle would always cross to the other side with 100% probabilty ($T = 1$), fully independent of the energy value $E > E_0$. However what we see is that the transmission probability goes through a set of maxima and minima, a phenomenon that is known in the literature as resonant transmission. It is clear that the maxima with $T = 1$ are induced by the zeros of the sine function: $k_2a = n\pi$, with $n = 1, 2, \cdots \infty$.

The particle wavelength λ *on* the barrier, given by the wave number $k_2 = 2\pi/\lambda$, is key to express the *resonance condition*, worth memorizing: *the barrier width must be an integer multiple of half the De Broglie wavelength*

$$a = n\frac{\lambda}{2} \qquad n = 1, 2, \cdots \infty .$$

What is remarkable of this phenomenon is not the fact that the particle crosses to the other side (as it occurs in Classical Mechanics), but the strong dependence on the energy that is found in the probability of transmission, depicted in Figure 25.2. A resonance occurs between the De Broglie wavelength and the dimensions of the obstacle that the particle passes through, provided that both quantities are of similar magnitude.

The most important cases happen when the obstacle is constituted by *individual atoms*, and the particle is an electron with energy in the eV range. Of course, if the obstacle repeats itself periodically (as in a crystal lattice), then the phenomenon is reinforced. Once the resonance condition is met for the first obstacle, it will be met for all of them, and the chain becomes transparent to the particle propagation.

The function (25.7), that is plotted in Figure 25.2, shows local maxima (resonances) associated with the integer values $n = 1, 2, \cdots \infty$, and also local minima when the argument of the sine function goes through multiples of $\pi/2$. It is not a universal function of E/E_0, but it depends critically on the particle *mass m* and the barrier width a.

Note the widening of the maxima with increasing values of n, while the transmission values $T \approx 1$ become less sensitive to the energy. This case actually corresponds to the classical limit.

In addition, for large values of n, it is the precision on our knowledge of the width a which is under test. Due to the intrinsic limitation on having a vertical potential step ($d \neq 0$), such high values of n are hardly observable, for any physical system.

These resonances are indeed met in the electric current conduction by solids and liquids, as a function of voltage (though not at the relatively high voltages chosen for transmission of electric power lines).

The historical Davisson-Germer experiment in 1927, that provided the first evidence of the De Broglie wavelength, may be considered the best example of the above idea of resonant transmission.

The Bragg law is fulfilled in this case: $n\lambda = 2d\sin\theta$, in a crystal lattice with the θ angle referring to the exfoliation surface of the crystal (glancing angle), that is nothing but the resonance condition applied to the interatomic distance d projected onto the electron path: $a = d\sin\theta$. This case is usually assessed as a two-dimensional problem, considering the electron as a wave propagating in the lattice. The detailed analysis reveals that the resonant wave is not actually transmitted, but *redirected* with an angle θ equal to the incidence angle, owing to the electron being simultaneously reflected by *all* parallel planes of atoms. But the phenomenon can indeed be regarded as a 1D resonant transmission case by a multiple barrier, the coordinate being taken along the electron path.

As far as the resonance on *individual* obstacles is concerned, we find the best example in the electric conduction by noble gases, due to the atom isolation in the gas, with rapidly falling E-field at the atom surface. Ramsauer and Townsend already observed in 1922 a clear resonance at 0.9 eV in the electric conduction by Xenon, that was a precursor to the early developments of Quantum Mechanics.

In all cases above, it is important to notice that formula (25.7) holds for *negative* values of E_0, which are those that properly approximate the electrical potential in 1D created by an individual atom (a being its diameter). The force on the electron is exerted only at its surface, and it is always *inward* directed. Likewise, on a very different length scale (fm), the above model describes the *nuclear* potential created by an atomic nucleus.

Despite the fact that our previous study has focused on the resonant transmission probability by the potential barrier with $E > E_0$ (D value in expression (25.1)), we may also be interested in the A coefficient (modulus and phase) that governs the reflected wave. It can easily be obtained, upon similar manipulation of the linear system (25.3). For the sake of completeness, we provide here the result:

$$A = \frac{\left(k_1^2 - k_2^2\right)\sin(k_2 a) A_0}{\left(k_1^2 + k_2^2\right)\sin(k_2 a) + 2ik_1 k_2 \cos(k_2 a)}.$$

26. QUANTUM TUNNELING

Let us now deal with the case where the energy of the particle incident on the potential barrier of width a, is *lower* than the barrier's height: $E < E_0$. As mentioned earlier, there is no essential difference with respect to the case $E > E_0$, from the mathematical standpoint. The wave number becomes *imaginary*, and we may call $\alpha \equiv ik_2 = \sqrt{2m(E_0 - E)}/\hbar > 0$ the positive constant of the real exponential, as we did for the step potential. Physics wise however, a fundamental difference arises, because the particle crossing to the region $x > a$ (with amplitude D) is now *classically forbidden*, the particle not having enough kinetic energy to overcome the barrier. As compared to the case $E < E_0$ for the step potential that we saw in Chapter 24, there is also a crucial difference: the particle may now retain its full kinetic energy in a permanent way, since the potential has dropped again to $U = 0$ at the distance a. This phenomenon is called quantum tunneling in the literature.

We must rewrite the general solution of $H\psi = E\psi$ (25.1) in the same three zones, using the parameter α above (and suppressing the subscript of the incident wave $k \equiv k_1$):

$$\psi_1 = A_0 e^{ikx} + A e^{-ikx} \qquad \psi_2 = B e^{-\alpha x} + C e^{\alpha x} \qquad \psi_3 = D e^{ikx}. \quad (26.1)$$

Similarly to the $E > E_0$ case, we now consider that the particle is incident from the left-hand side as a boundary condition. According to the physical interpretation of the coefficients of the plane waves discussed earlier, the transmission probability is to be determined as:

$$T = \frac{|D|^2}{|A_0|^2} = \frac{DD^*}{|A_0|^2}.$$

Chapter 26. QUANTUM TUNNELING

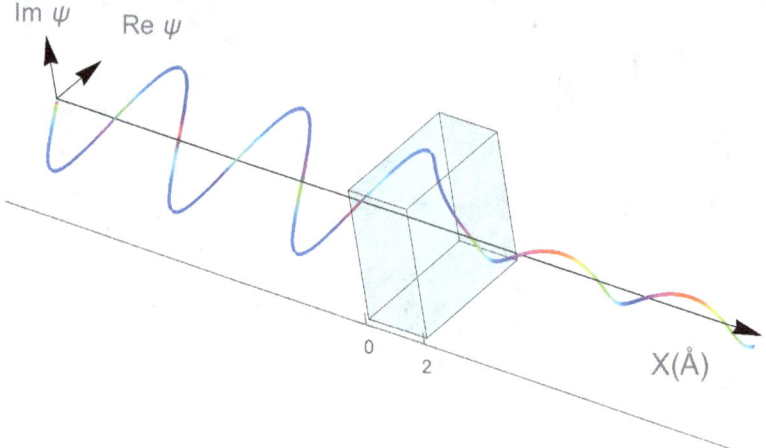

Figure 26.1: *Wave function of an electron with $E = 8\,eV$ that tunnels past a potential barrier of width $a = 2\,Å$ and height $E_0 = 10\,eV$. The wave function has been represented in the complex plane $(Re\,\psi, Im\,\psi)$, as function of the distance x travelled by the electron. The function $\psi_2(x)$ inside the barrier is constrained by the 4 continuity points of ψ and $\partial \psi/\partial x$ at $x = 0$ and $x = a$, analytically determined by the linear system (26.2). The tunneling probability is $T = 0.136$. The color code evolves with the phase. Note its nearly constant value on the left-hand side, where the function is almost real, as a consequence of the intense reflected wave.*

For clarity, let us write again the equations that express the continuity of ψ and of $\partial \psi/\partial x$ at the points $x = 0$ and $x = a$:

$$A_0 + A = B + C \qquad ikA_0 - ikA = -\alpha B + \alpha C$$
$$Be^{-\alpha a} + Ce^{\alpha a} = De^{ika} \qquad -\alpha Be^{-\alpha a} + \alpha Ce^{\alpha a} = ikDe^{ika} \qquad (26.2)$$

By matrix inversion of the 4×4 linear system we can uniquely determine the complex constants $A, B, C, D \in \mathbb{C}$. The unique solution found for the wave function in the three zones (26.1) has been represented in Figure 26.1, for a tunneling electron under conditions given therein. More handily, the coefficients can be obtained by repeating the steps performed for the equations with $E > E_0$. If we are interested in the D coefficient, that determines the probability of transmission, we arrive without difficulty to the expression:

$$-4ik\alpha A_0 = \left[(k-i\alpha)^2 e^{-\alpha a} - (k+i\alpha)^2 e^{\alpha a}\right] De^{ika} \,. \qquad (26.3)$$

We also provide, for completeness, the expression obtained for A:

$$A = \frac{(k^2 + \alpha^2)\sinh(\alpha a)A_0}{(k^2 - \alpha^2)\sinh(\alpha a) - 2ik\alpha\cosh(\alpha a)}.$$

It is no surprise that in passing from $E > E_0$ to $E < E_0$ we must swap the sine and cosine functions for the *hyperbolic sine* and the *hyperbolic cosine*, respectively, as a consequence of the relation $\alpha \equiv ik_2$ [1]. The final expression that is found for the tunneling probability is the following:

$$T = \frac{1}{1 + \dfrac{\sinh^2(\alpha a)}{4\dfrac{E}{E_0}\left(1 - \dfrac{E}{E_0}\right)}}. \tag{26.4}$$

It is governed by the dimensionless quantity αa. For E/E_0 not too close to zero ($E \not\to 0$), and not too close to one ($E \not\to E_0$), conditions that are met in all cases of practical interest, the tunneling probability can be obtained as a simple exponential:

$$T \simeq 16 \frac{E}{E_0}\left(1 - \frac{E}{E_0}\right) e^{-2\alpha a} \simeq e^{-2\alpha a}.$$

The real quantity $\hbar\alpha > 0$ is often called in the literature imaginary momentum, because it represents the modulus of the imaginary value of the momentum that the particle has inside the barrier (with negative kinetic energy). The approximation $T \sim e^{-2\alpha a}$ offers a way to calculate the tunneling probability through a potential barrier whose line shape may not be rectangular, but arbitrary, as defined by the function $U(x)$.

Just imagine, in this case, that the particle goes through a multiplicity of very narrow rectangular barriers (of width dx), at successive locations $(x, x+dx)$. It is natural to associate the probability of crossing N successive barriers with the *product* of the probabilities of each of them:

$$T \simeq e^{-2\alpha_1(x)dx} \cdot e^{-2\alpha_2(x)dx} \cdots = e^{-2\left(\alpha_1(x) + \alpha_2(x) + \cdots\right)dx}. \tag{26.5}$$

Let us designate by x_1 and x_2 the points where the barrier begins and ends, respectively, as defined by the equation $U(x) = E$, depicted in Figure 26.2. These two points are called *return points* in Classical Mechanics, where a zero velocity is crudely attributed to the particle (in fact, such velocity is never realized).

---

[1] be reminded of the properties $i\sin(ix) = \sinh x$ and $\cos(ix) = \cosh x$.

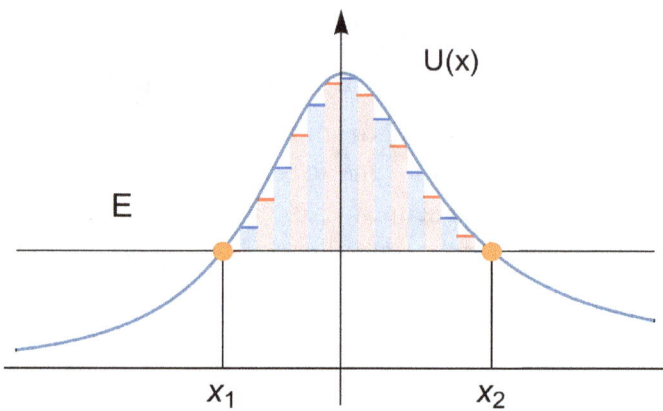

Figure 26.2: *Return points (x_1,x_2) for the calculation of the tunneling probability through a potential barrier with line shape given by the function $U(x)$, with E being the energy of the particle. The barrier is approximated by a multiplicity of very narrow rectangular ones, whereby the Riemann integral of the imaginary momentum is used.*

If we divide the interval $x_2 - x_1$ in N equal subintervals having width $dx = (x_2 - x_1)/N$, then the exponent above corresponds to the *Riemann integral* of the function $\alpha(x)$ in the interval (x_1,x_2), in the limit $N \to \infty$. This function tells us the imaginary momentum at each point. Thus the tunneling probability for a particle of total energy E, through the barrier defined by $U(x)$, can be approximated by:

$$T \simeq e^{-\frac{2}{\hbar}\int_{x_1}^{x_2}\sqrt{2m(U(x)-E)}dx} . \qquad (26.6)$$

It is interesting to recognize in the integral over the imaginary momentum the *reduced action* for the particle propagation over a formally classical trajectory through the barrier: $S_0 = \int_{x_1}^{x_2} p(x)dx = i\int_{x_1}^{x_2} \hbar\alpha(x)dx$. What has happened is that the action has become imaginary. As a result, the Feynman propagation amplitude for this trajectory is now real ($i \cdot i = -1$). Of course, we must associate the tunneling probability with the *squared modulus* of the amplitude.

Hence, we have:

$$T \simeq |A|^2 \left|e^{-iEt/\hbar}\right|^2 \left|e^{iS_0/\hbar}\right|^2 = |A|^2 e^{-2|S_0|/\hbar} ,$$

that correctly reproduces the exponential expression (26.6), directly deduced now from the principles of Quantum Mechanics, without the long derivation using Schrödinger's equation previously discussed.

Note the factor 2 in the exponent (critical for any numerical application), is rooted in the probabilistic interpretation of the *square* of the wave function $|\psi|^2$. Of course, this includes the particular case of the rectangular barrier:

$$T \simeq e^{-2\alpha a}.$$

It is also understood why it is an *approximate* calculation: the integration is performed over the "classical" trajectory of imaginary velocity (the closest to Classical Mechanics with constant energy E), which is only the *dominant one* within the full path integral (3.5) that we saw in Section 3.1.

In summary, what quantum tunneling means is that a particle incident with energy E on a potential barrier that is impenetrable in Classical Mechanics ($E < E_0$) may be observed on the other side of the barrier, with a calculable probability. Note that, as part of the calculation leading to the tunneling probability (26.4), we have used (in one case) the stationary states of the Schrödinger equation, with well defined energy ($\Delta E = 0$), despite the fact that these are deprived of any time resolution ($\Delta t = \infty$).

Let us recall that the condition of incidence from the left was stated in the calculation as a *spatial* (and *complex*) boundary condition on the differential equation $H\psi = E\psi$. The stationary states actually represent the *ideal* limiting case of an incident particle whose wave train is infinitely long.

What we actually have tipically in the laboratory is something similar to three cylindrical capacitors as those represented in Figure 24.1, where the last one has been grounded to zero potential with the first one. The particle may produce an electronic signal that triggers the start of a clock inside the first capacitor and another signal that hypothetically triggers the clock stop ($\Delta t \neq \infty$), *after* tunneling has occured. The stationary state has allowed us to mathematically *approach* this situation, assuming for simplicity that $\Delta t \to \infty$.

The time evolution of an incident *packet* with $\Delta t \neq \infty$ may also be treated by means of the full Schrödinger equation $i\hbar\, \partial \psi/\partial t = H\psi$, through analysis of its inverse Fourier transform. We are not doing it here, but rather describe the result: the incident packet is *distorted* upon arrival to the barrier, and it *splits* into two, that move apart from each other. One of them bounces back in the opposite direction, and the other one moves forward, with a reduced amplitude. Yet it is obvious that the particle *cannot be divided* as its wave function is, and the latter must collapse at some point.

A *transmission time* δt can be associated to the tunneling process, based upon the above-mentioned analysis of the Fourier integral, for a gaussian packet of width ΔE [2]. The method consists of requiring the phase of the transmitted wave (c-number D in our previous analysis of the rectangular barrier (26.3)) to be an *extremum*, as for the Feynman integral in Chapter 4. It yields, for $E, \Delta E \ll E_0$ and not too thin barriers, a remarkably short time: $\delta t \approx \hbar/\sqrt{EE_0}$.

We shall outline four key features of quantum tunneling:

- The *strict energy conservation*. The particle that crosses to the other side retains 100% of its total energy E. A popular description of tunneling is saying that the particle "borrows" the necessary energy ΔE to overcome the barrier from quantum fluctuations, during a short period of time Δt, according to the uncertainty principle $\Delta E \Delta t \sim \hbar$. Past this time, the energy is "given back" and its original value restored. This image is indeed useful, although it cannot describe at all the case approaching a stationary state, where $\Delta E = 0$ by construction.

- Its perfectly *relativistic* character, as being originated from the De Broglie wave with a real exponent, despite having used the non relativistic form of the imaginary momentum. Using a relativistic expression for the imaginary momentum in the integral would entirely make sense.

- Its characteristic *exponential* dependence on the product of twice the imaginary momentum times the barrier width ($2\alpha a$). At equal distance, a strong suppression is seen with increasing particle mass m (classical limit). For electrons, tunneling typically takes place at the nanometer distance scale, whereas for nucleons, it is more likely to happen at the femtometer scale.

- Its *non time reversible* character. No solutions exist for an electron coming back from the rear side without reflection, increasing its amplitude through the barrier, and emerging at the front side as an almost real wave (as embodied by time running backward $t \to -t$, see Figure 26.1). The only existing symmetrical solution is represented by opposite side incidence with equal transmission (26.4). Which means not only $t \to -t$, but also $\psi \to \psi^*$, in order to verify $i\hbar \partial \psi/\partial t = H\psi$.

---

[2] T. E. Hartman, Journal of Applied Physics **33**, 3427 (1962). To be apprised of the method, and aware of its implication on very high frequency diodes, see the dedicated exercise of the course.

Quantum tunneling plays a central role in numerous technological devices where its quantum basis is well established, that we shall comment below. Some of them meet frontier developments, such as controlled nuclear fusion. It also plays an important role in the field of molecular biochemistry, in relation, among other topics, with photosynthesis and cellular respiration. Its key role has also been pointed out in the primordial synthesis of important prebiotic molecules in astrochemistry [3].

As an example of the well known topics of quantum tunneling in physics we shall quote the following, that are dealt with, at introductory level, in specific exercises of the course:

- The emission of α particles. For its importance, we shall discuss it here in some detail, as it can serve as a model for all the other cases. Clarified by G. Gamow in 1928, it is one of the historical highlights of Quantum Mechanics.

- The scanning tunneling microscope. Free electrons in a metal, that are trapped by the potential step at its surface (the workfunction w_0), can be extracted on application of a strong external electric field. It allows to observe, with clear definition, the size and clustering of the atoms on the metal surface.

- The field emission diodes. They allow commutation of electrical currents at very high frequencies in some semiconductors, since they are not based on the diffusion process (as normal diodes are), but in the quantum frequency of De Broglie. This is an ample field, with many implications in very large scale integration (VLSI) technologies, applied, for example, to mobile devices.

- The oscillacion of the NH_3 molecule. The Nitrogen atom could not classically pass through the triangle formed by the Hydrogen atoms, but it does, and comes back at a very stable frequency ω. Two potential wells (with classical frequency ω_0) become connected by a potential barrier with tunneling probability T. The periodic transit motion of the N atom arises from the appearance of two very close energy levels, that raise *Bohr oscillations* with: $E_2 - E_1 = \hbar\omega = \hbar(T\omega_0)$.

---

[3] see for example the article "Quantum Tunnelling to the Origin and Evolution of Life", F. Trixler (2013), https://www.ncbi.nlm.nih.gov/pubmed/24039543

- The `emission of electron-positron pairs`, a phenomenon that admits an interpretation as quantum tunneling, where the electrons are extracted not from a metal, as in the case of the tunneling microscope, but from a vacuum. The strong electric field can be provided, for instance, by Coulomb's law exerted on the surface of an atomic nucleus. The probability calculation can actually be performed in an identical way as for the tunneling microscope, in a non relativistic approach. The positron is interpreted as the *hole* left by the electron in the vacuum.

- The phenomenon of `nuclear fusion`, related with that of `stellar nucleosynthesis`, that governs the primordial production of heavy nuclei in high mass stars. It should be noted that stars themselves would be unsustainable unless for tunneling, since the electric repulsion between nuclei would preclude the fusion process. As to the formalism, it is similar to that of α particle emission, since the light nuclei need to overcome the same Coulomb barrier, yet in opposite direction.

The lifetime of α-emitters

Many heavy nuclei are α [4] particle emitters, as part of the phenomenon of natural radioactivity. Circumventing a more detailed treatement in nuclear physics, it makes sense to consider that the α particle preexists inside the atomic nucleus before the emission, particularly in the case of emitters with even number of nucleons.

Each α emitter isotope is characterized by two quantities, measurable with high precision: its lifetime τ and the kinetic energy E_α of the α particle. While the former show a wide range of variation, from the microsecond to thousands of years, the values of E_α lie in the range of tens of MeV, not far from the energies they actually have inside the nucleus.

The *nuclear activity* (per mole) is defined as $\lambda_N = N_A \lambda$, with $\lambda \equiv 1/\tau$. Our reference variable from now on will be λ (that is measured in s^{-1}). When the lifetime is very short, one mole of emitter has an enormous nuclear activity, i.e. it emits many α particles per minute (thus becoming highly hazardous).

---

[4] the α particle is the nucleus of a Helium atom, and it is formed by two protons and two neutrons.

Large variations in the activity of α emitters were already observed over the decade of 1920s, a time when the energies E_α were also measured, mostly thanks to the pioneering work by Gieger and Nutall [5].

These authors had established an empirical relationship between the above quantities, that consisted in a linear dependence between the logarithm of λ and $1/\sqrt{E_\alpha}$, the so-called Geiger-Nutall law, but this law could not be understood with the physics known at the time. In Figure 26.3 we have collected current data on α emitters that show excellent agreement with Geiger-Nutall's law, for all even isotopes of Uranium, Polonium, and Radium, to be analysed below.

Not only was the Geiger-Nutall law not understood, but when the newer data on *nuclear radii R* (in *fm*) were incorporated, physicists were puzzled by a paradox that seemed unsolvable by the laws of mechanics, and in conflict with the principle of energy conservation. These measurements of R were achieved following the method introduced by Rutherford to scatter precisely α particles off nuclei at different energies. The key observation was the limiting (maximum) energy at which these were absorbed by the nucleus. It is easily seen that the impact parameter in this case must equal the nuclear radius R.

The above-mentioned paradox can be grasped if we take the example of the ^{238}U isotope data, with a radius $R \approx 30\,fm$ [6] and the measured value $E_\alpha = 4.2\,\text{MeV}$. Coulomb's law provides the potential energy of the α particle outside the nucleus, at a distance $r \geq R$ from its center:

$$U(r) = \frac{1}{4\pi\varepsilon_0}\frac{2Ze^2}{r} .$$

Taking $Z = 92$ for Uranium, the potential energy on the nucleus surface at $r = R = 30\,fm$ can be evaluated. The factor 2 accounts for the charge of the α particle. The result is $B \equiv U(R) = 8.8\,\text{MeV}$, as shown in Figure 26.4.

---

[5] H. Geiger carried out in 1909 with E. Marsden the famous gold foil experiment that revealed the existence of the atomic nucleus, and invented the counter named after him, that keeps being today the reference device used for α radioactive detection. The measurement of E_α can be achieved by different methods, the most common being its energy loss when slowed down in matter. Precisions of $\sim 70\,KeV$ can be attained today.

[6] nuclear radii are not exempt from theoretical uncertainty, no matter the accuracy of experimental measurements, due to the fact that the nuclear potential does not increase to its maximum value sharply, but over a certain distance, determined by the details of the electric field on the nucleus surface. Recall the conceptual discussion of the cylindrical capacitor in Figure 24.1, now with $d \sim 5\,fm$. Nevertheless the above-named paradox is perfectly well illustrated by taking the *average* value of $30\,fm$, as indicated.

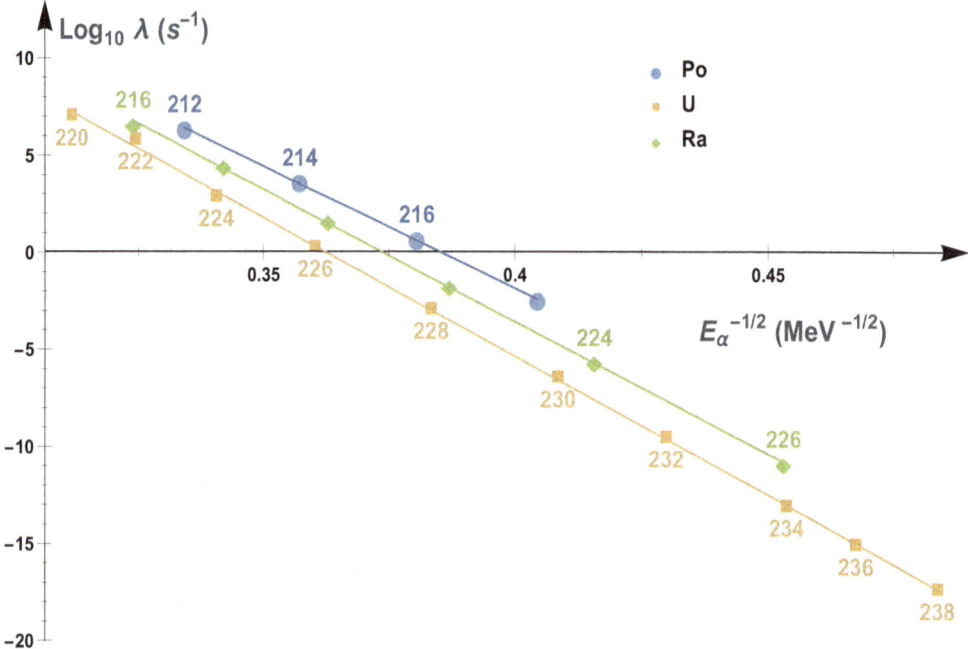

Figure 26.3: *Nuclear activity λ of the even α emitting isotopes of Uranium (U), Polonium (Po) and Radium (Ra), as function of the inverse of the square root of E_α. The observed linear dependence on each series is in excellent agreement with the Geiger-Nutall law discussed in the text. Note the 25 orders of magnitude variation on the value of λ. Source: Isotope Data, Wolfram Mathematica 11.*

How to explain that the measured kinetic energy of the α particle is *only* 4.2 MeV? The principle of energy conservation tells us that if the particle has gone through the nucleus surface, it necessarily acquires an energy of 8.8 MeV that will be converted into kinetic energy, no matter the trajectory followed by the α particle until the Geiger counter is hit, which is located at zero potential.

In 1928 the Russian-American physicist G. Gamow brilliantly explained the above paradox, as well as the Geiger-Nutall law itself, by using the Schrödinger equation. This led him to introduce the idea of quantum tunneling, to which he gave birth, just in the analytical manner we have discussed earlier, for a potential barrier.

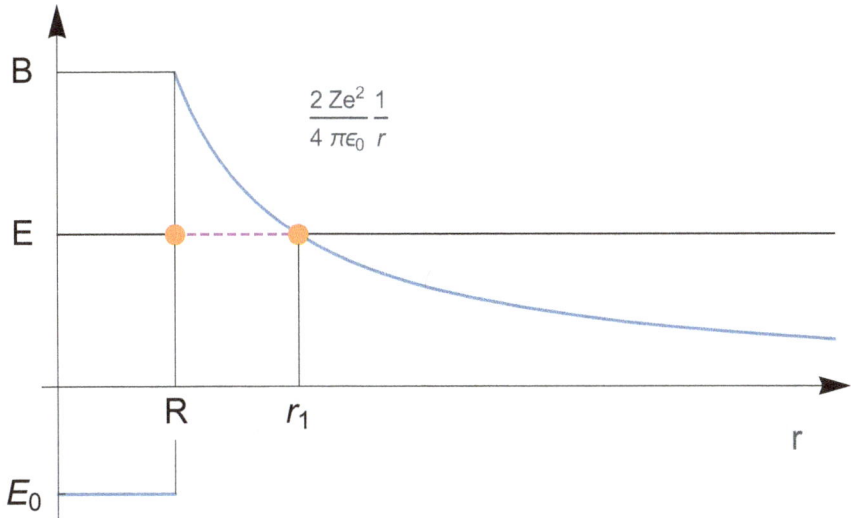

Figure 26.4: *Potential energy of an α-particle inside and outside an ideal nucleus of radius R, as function of its distance to the center. Inside the nucleus, it is due to the nuclear force, and it is constant. Outside, it originates from Coulomb repulsion. The α-particle may escape by quantum tunneling, keeping its energy E constant. The parameters B and r_1 refered to in the text are also indicated, along with the energy values $E = 4.2\,MeV$ and $B = 8.8\,MeV$, as they are realized for ^{238}U. A more realistic model would show a finite potential slope near R.*

Gamow chose the name *tunneling* motivated by a gravitational version of the barrier, embodied in an ascending mountain road. A car left to its own inertia, with insufficient kinetic energy $mv^2/2$, could never get past a certain height, and would then return along the same path in opposite direction, to regain its original velocity (-v). However, quantum physics allows it to go down *the other side* of the mountain (without ever getting to the top), and equally regain its original velocity (v), *as if a tunnel had been dug* spontaneously across the mountain.

Let us now see in detail Gamow's derivation of the Geiger-Nutall law from the Coulomb barrier seen by the α particle in a nucleus, as represented in Figure 26.4. It is a key idea to realize that inside the nucleus the α particle is subject to a constant potential, with a *negative* value $-|E_0|$ (in MeV), with respect to the zero potential defined at infinite distance. The inner potential drops to such negative level (specific to each isotope), due to the intense nuclear attraction force, that overcomes the electrostatic repulsion between the protons.

Within the nucleus, it is then a good approximation to consider that the α particle moves with constant velocity v_0, and bounces back when it hits the surface from the inside, with frequency $v_0/(2R)$ (two times per cycle). Thus the nuclear activity of the emitter is well approximated by the expression:

$$\lambda = \frac{v_0}{2R} e^{-2G}, \tag{26.7}$$

where G is the integral of the imaginary momentum (divided by $\hbar$) between the two return points indicated in Figure 26.4:

$$G = \frac{1}{\hbar} \int_R^{r_1} \sqrt{2m(U(r) - E_\alpha)} dr.$$

The return points are calculated as the intersection of the horizontal line $E_\alpha = const$ with the barrier, that implies the proportion: $x \equiv R/r_1 = E_\alpha/B$. Recall the barrier's top height:

$$B = \frac{1}{4\pi\varepsilon_0} \frac{2Ze^2}{R}.$$

The integral can be calculated analytically after a simple change of variable [7], and we directly write down the result, already simplified:

$$G = \frac{(2m)^{1/2}}{\hbar} \int_R^{r_1} \sqrt{\left(\frac{1}{4\pi\varepsilon_0}\frac{2Ze^2}{r} - E_\alpha\right)} dr = \frac{1}{\hbar}\left(\frac{2m}{E_\alpha}\right)^{1/2}\left(\frac{2Ze^2}{4\pi\varepsilon_0}\right)\left[\frac{\pi}{2} - \arcsin\sqrt{x} - \sqrt{x(1-x)}\right]$$

Taking into account that the dimensionless quantity x defined above is small with respect to one for every emitter, it makes sense to perform the power series expansion of the expression in square brackets. Note that, in addition, the factors in front of it include the inverse of the *velocity* of the α particle: $v_\alpha = \sqrt{2E_\alpha/m}$. That is, exactly what appears in the Geiger-Nutall law. The expression in square brackets $\gamma(x)$ includes an arcsine function, and taking the first two terms of its series expansion in powers of x is generally sufficient, the first one being dominant

$$\gamma(x) = \frac{\pi}{2} - 2\sqrt{x} + \cdots,$$

that leads us to the final expression:

$$G \simeq \frac{Ze^2}{2\varepsilon_0 \hbar v_\alpha} - \frac{2e}{\hbar}\sqrt{\frac{ZRm}{\pi\varepsilon_0}}.$$

---

[7] the change of variable is: $r = r_1 \sin^2\theta$. For $r = R$ we have $\theta = \arcsin(R/r_1)^{1/2}$, and for $r = r_1$, we have $\theta = \pi/2$. In the final integral $\cos^2\theta = (1+\cos(2\theta))/2$ should be taken into account.

Introducing now the nuclear activity of the emitter through expression (26.7), the Geiger-Nutall law is proven, in the form:

$$\ln \lambda = -\frac{Ze^2}{\varepsilon_0 \hbar}\sqrt{\frac{m}{2E_\alpha}} + \ln\left(\frac{v_0}{2R}\right) + \frac{4e}{\hbar}\sqrt{\frac{ZRm}{\pi\varepsilon_0}} = -KE_\alpha^{-1/2} + C\ .$$

Where we see that the slope K of the linear dependence of $\ln\lambda$ with $E_\alpha^{-1/2}$ only depends on Z. This is precisely the reason for taking different isotopic series of α emitters for a given element Z, as we have done in Figure 26.3 with Uranium (U), Polonium (Po), and Radium (Ra). The intercept C additionally depends on v_0 and R, two parameters that are fundamentally determined by the Z value, as shown by the results in Figure 26.3 themselves, and may change in general when going to a different Z.

Note the v_0 constant represents the velocity of the α particle *inside* the nucleus, that compiles the information of the nuclear potential $-|E_0|$ represented in Figure 26.4. This velocity is in general *different* from v_α (the measured velocity at the Geiger counter), with the relationship: $(1/2)mv_0^2 = (1/2)mv_\alpha^2 + |E_0|$.

The excellent agreement that is observed between the data and the Gamow hypothesis of quantum tunneling, *spanning 25 orders of magnitude*, constituted at the time, and still constitutes today, a stunning confirmation of Quantum Mechanics. There is no other physical law, as known today, that keeps agreement with the data over such a long range of variation of its measurable parameters (think for instance of electromagnetism, Einstein's gravitation law, etc.)

What has been learned from the analysis of the above paradox, as to the unequivocally non local, interpretation of Quantum Mechanics, is not to be minimized. It must be admitted that the α particle resides *simultaneously* inside the atomic nucleus and in the vicinity of the Geiger counter, while no signal has been registered in the latter.

27. THE 3D HARMONIC OSCILLATOR

Quantization of the reduced classical action allowed us to obtain in Section 2.4 the energy spectrum of the harmonic oscillator. Let us recall that it was precisely for the harmonic oscillator that Planck introduced in 1900 the h constant, and conjectured its possible energy discretization. Its profound impact on the physics of the early 20th century has already been pointed out. We shall now deal with the problem of solving Schrödinger's equation $H\psi = E\psi$ to find out the exact solution for the stationary states, and compare. But more importantly, determine the wave functions ψ_n of each energy level, and analyze their 3D extension, and quantum degeneracy.

We take the potential energy in 1D as: $U(x) = (1/2)m\omega^2 x^2$ (where $\omega = \sqrt{k/m}$ is the oscillator's characteristic frequency), with the natural choice $U(0) = 0$, where only positive energies $E > 0$ make sense. Recall that the amplitude of the classical solutions is uniquely determined by the energy: $A_{cl} = (1/\omega)\sqrt{2E/m}$.

The differential equation $H\psi = E\psi$ is written in the form:

$$\frac{-\hbar^2}{2m}\frac{d^2\psi}{dx^2} + \frac{1}{2}m\omega^2 x^2 \psi = E\psi, \qquad (27.1)$$

which, after defining $\alpha \equiv m\omega/\hbar$ and $\beta \equiv 2mE/\hbar^2$, becomes:

$$\frac{d^2\psi}{dx^2} + \left(\beta - \alpha^2 x^2\right)\psi = 0.$$

Let us introduce a new dimensionless variable u to measure the elongation of the oscillator, as:

$$u \equiv \sqrt{\alpha}x = \sqrt{\frac{m\omega}{\hbar}}x,$$

with which the equation takes the new form:

$$\frac{d^2\psi}{du^2} + \left(\frac{\beta}{\alpha} - u^2\right)\psi = 0.$$

Without loss of generality, we may now perform a change of function, and solve for a new function $H(u)$, defined as:

$$\psi(u) \equiv e^{-u^2/2}H(u), \qquad (27.2)$$

so once a solution is obtained for $H(u)$, equation (27.2) is used to retreive the wave function $\psi(u)$. By using the chain rule, one arrives straightforwardly to the new equation for $H(u)$:

$$\frac{d^2H}{du^2} - 2u\frac{dH}{du} + \left(\frac{\beta}{\alpha} - 1\right)H = 0.$$

As done in previous cases, we now use the method of solving the equation by performing the power series expansion of $H(u)$ around $u = 0$, and establishing a recursive relation on its coefficients. Presumably, a convergence radius $(-\infty, +\infty)$ shall be obtained, as expected for the quantum oscillator wave function. Indeed, let us recall the idea of Feynman's propagator, that will certainly enable the oscillator to break the limits of Classical Mechanics and propagate beyond its classical amplitude A_{cl}.

The key question is whether the solutions of the differential equation for $\psi(u)$ are *square-integrable* for any choice of the energy E. To investigate that, we write down the expansion:

$$H(u) = a_0 + a_1 u + a_2 u^2 + \cdots = \sum_{j=0}^{\infty} a_j u^j,$$

and establish now the recurrence law on its coefficients a_j that is obtained by summing term by term all contributions proportional to H'', H' and H, and then grouping all terms together. We omit the grouping task and simply indicate the result that is obtained:

$$a_{j+2} = -\frac{(\beta/\alpha) - 1 - 2j}{(j+1)(j+2)} a_j \qquad j = 0, 1, \cdots \infty. \qquad (27.3)$$

It can be shown, on application of the ratio test, that indeed the radius of convergence of the above series is $(-\infty, \infty)$. It is important to note that the recurrence law (27.3) is valid for both the *even* and *odd* terms of the series. Which means that, once the a_0 and a_1 coefficients are fixed, all the other terms are determined. Hence these numbers are the two arbitrary integration constants that every second order differential equation must have.

Now we focus on the asymptotic behaviour of $H(u)$ for $u \to \pm\infty$ (the boundary of the convergence region), by application of the Lemma of Asymptotic Similarity we stated at the beginning of Chapter 16. The reference series, that actually justifies our previous change of function, is the exponential function itself:

$$e^{u^2} = 1 + u^2 + \frac{u^4}{2!} + \frac{u^6}{3!} + \cdots.$$

Using the symbol $\sim$ to denote asymptotic equality at the indicated limit, the recursive law is, obviously:

$$\frac{a_{j+2}}{a_j} = \frac{(j/2)!}{(j/2+1)!} = \frac{2}{j+2} \sim \frac{2}{j} \quad \text{for } j \to \infty.$$

It is clear that, for very large values of j, the recurrence law (27.3) approaches the limit: $a_{j+2}/a_j \sim 2/j$, which is asymptotically equivalent to the above, thus the function $H(u)$ does not verify $H(u) \to 0$ for $u \to \pm\infty$, a necessary condition to be square-integrable. Note the same reasoning applies separately for each of the above series: the even and the odd terms. One needs to check now whether the necessary condition is still verified by the *wave function* itself, as defined by (27.2). But this is not the case either, due to the factor 2 then introduced in the exponent:

$$\psi(u) \equiv e^{-u^2/2} H(u) \sim A_2 e^{u^2/2} + A_1 e^{u^2/2}, \tag{27.4}$$

where we have taken into account that the series $H(u)$ is the sum of two (even and odd terms), both divergent for $u \to \pm\infty$ by virtue of the Lemma of asymptotic similarity. The coefficients $A_{2,1}$ are proportional to $a_{0,1}$ (multiplied by two constants, as stated in the Lemma).

Table 27.1: Hermite polynomials up to $n = 5$

$H_0 = 1$	$H_1 = 2x$
$H_2 = 4x^2 - 2$	$H_3 = 8x^3 - 12x$
$H_4 = 16x^4 - 48x^2 + 12$	$H_5 = 32x^5 - 160x^3 + 120x$

Table 27.2: Wave functions of the 1D harmonic oscillator ψ_n up to $n = 3$

$\psi_0(x) = (a/\sqrt{\pi})^{1/2} e^{-a^2 x^2/2}$	$\psi_1(x) = (a/2\sqrt{\pi})^{1/2} 2ax e^{-a^2 x^2/2}$
$\psi_2(x) = (a/8\sqrt{\pi})^{1/2} (4a^2 x^2 - 2) e^{-a^2 x^2/2}$	$\psi_3(x) = (a/48\sqrt{\pi})^{1/2} (8a^3 x^3 - 12ax) e^{-a^2 x^2/2}$

Thus we have arrived to the surprising conclusion that no solution to equation (27.1) provides a square-integrable wave function ψ. This would mean a dead end to the Schrödinger equation, and the only way out is realizing that, for some critical values of E, the series is not actually formed by an infinite number of unequal terms. Instead, all of them become zero after a given index j. If we look carefully at the numerator of expression (27.3), we see what the condition is for this to happen: the ratio $\beta/\alpha \in \mathbb{R}$ must be an *odd integer*.

That is, explicitely: $\beta/\alpha = 2n + 1$, with $n = 0, 1, \infty$. On substitution of the previously defined parameters β (energy) and α, we get the renowned formula for the energy quantization of the 1D harmonic oscillator:

$$E_n = \left(n + \frac{1}{2}\right)\hbar\omega \qquad n = 0, 1, \cdots \infty, \qquad (27.5)$$

thus confirming that the semiclassical expression (with Maslov index $\mathcal{M} = 2$) actually provides the *exact* result that is obtained from Schrödinger's equation. It is not extremely surprising after all, that quantum fluctuations taking place symmetrically around the classical trajectories, have a tendency to compensate and leave propagation over the classical trajectory unaltered.

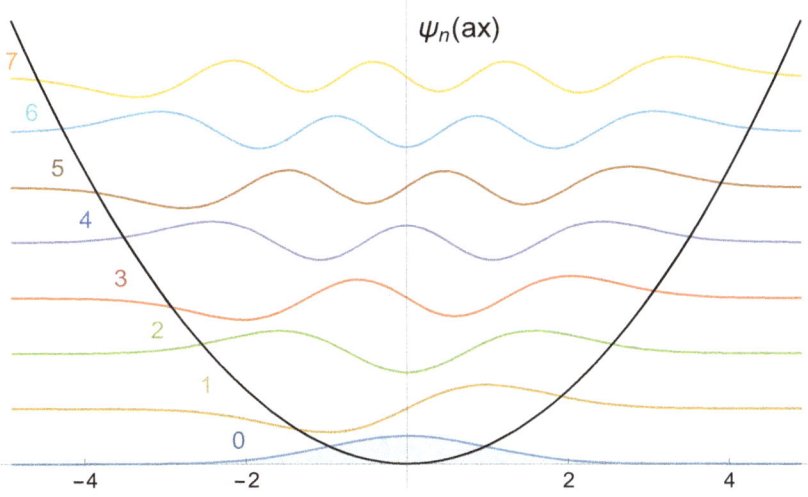

Figure 27.1: *Wave functions $\psi_n(ax)$ of the stationary states of the 1D harmonic oscillator, from $n = 0$ to $n = 7$. Note the alternating parity of the wave functions. For the sake of clarity, and illustration of their energy spacing, they have been displaced on the vertical scale by a fixed constant. The horizontal scale is expressed in units of the characteristic oscillator length $1/a = \sqrt{\hbar/(m\omega)}$. The area of the ground state has been filled in blue. The black parabola represents the potential energy.*

We shall now examine the wave functions $\psi_n(x)$ of each stationary state of energy E_n. The structure of the solutions is already indicated by equation (27.2): the product of a polynomial times a gaussian function. Yet the polynomial only contains even or odd powers. Indeed, condition $\beta/\alpha = 2n + 1$ only garantees that one of the two series (even or odd terms) degenerates into a polynomial, in order to render $\psi(u) \to 0$ at infinity from equation (27.4). The other needs to be set to zero by the integration constant $A_{1,2} = 0$, depending on the parity of n.

The polynomials meeting odd-integer values of β/α in the recurrence law (27.3) are well documented in the literature as *Hermite polynomials*. They can be evaluated in a recursive manner themselves by means of the Rodrigues formula:

$$H_n(x) = (-1)^n e^{x^2} \frac{d^n}{dx^n} e^{-x^2}.$$

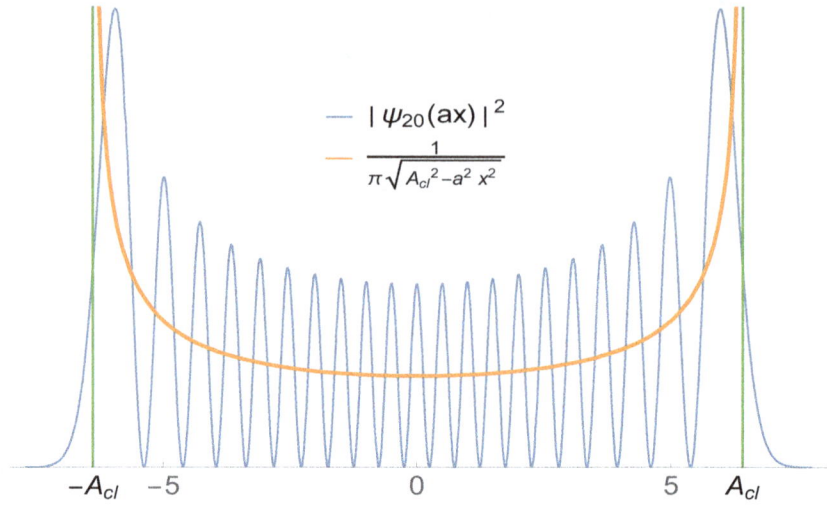

Figure 27.2: Probability density $|\psi_n(ax)|^2$ of a harmonic oscillator with $n=20$, as a function of its elongation. The classical probability density, that would be shown by random pictures of the oscillator, has been superimposed in orange color. Note the latter becomes infinite at the classical amplitude A_{cl}, where Classical Mechanics crudely assumes the velocity to be zero. Yet the agreement between both curves gets better and better as $n \to \infty$, given that a limited position resolution will always result in observing the average probability.

We provide in Table 27.1 all polynomials up to $n=5$. Given that the previously defined constant $a \equiv \sqrt{\alpha} = \sqrt{m\omega/\hbar}$ is measured in m^{-1}, the wave functions actually depend on the dimensionless argument ax, and can be expressed as:

$$\psi_n(ax) = N_n H_n(ax) e^{-a^2 x^2/2},$$

where the constant N_n arises from the normalization condition:

$$\int_{-\infty}^{\infty} N_n^2 H_n^2(ax) e^{-a^2 x^2} dx = 1,$$

leading to the value:

$$N_n = \sqrt{\frac{a}{\sqrt{\pi} 2^n n!}}.$$

The wave functions of the lowest states are given in Table 27.2, and have been depicted in Figure 27.1. Note that the gaussian function factorizing all states has the width $\sigma = 1/a = \sqrt{\hbar/(m\omega)}$, that provides the characteristic length of quantum fluctuations for every harmonic oscillator, that is the exact analog to the Bohr radius of the atoms.

We see that the ground state is a simple gaussian, with a local maximum of the probability density at the equilibrium position. This behaviour is actually *opposite* to that of Classical Mechanics. Indeed, the latter also allows to talk about a probability density to find the oscillator at a given point, just by making random pictures of its position. The probability would then be inversely proportional to the oscillator's velocity at each point, that is maximal at the origin $x = 0$.

It is remarkable that the width σ of the ground state gaussian precisely corresponds to the classical value of the amplitude, taken with the ground state energy $E_0 = (1/2)\hbar\omega$, $\sigma = A_{cl} = \sqrt{2E_0/m}(1/\omega)$. This was to be expected, since the maximum violation of Classical Mechanics naturally takes place at the ground state. Another way to say this is that the oscillator spends 32% $(1 - P_{1\sigma})$ of its time outside the classical amplitude.

Even more remarkable is the fact that, for large values of n, the quantum distribution of the probability density is amazingly close to its classical version, reaching an *absolute maximum* around $x = A_{cl}$, for $n \to \infty$, as it can be appreciated in detail in Figure 27.2, already for $n = 20$.

Hermite polynomials satisfy the following orthogonality relations:

$$|N_n|^2 \int_{-\infty}^{+\infty} H_{n'}(x) H_n(x) e^{-a^2 x^2} dx = \delta_{n'n}, \qquad (27.6)$$

that ensure the orthonormality of the states: $\langle \psi_{n'} | \psi_n \rangle = \delta_{n'n}$.

Because the Hermite polynomials only have even or odd powers, and being the gaussian an even function, the $\psi_n(x)$ states have a well defined parity , i.e.: $\psi_n(-x) = (-1)^n \psi_n(x)$. The following recursive relation for the polynomials themselves, well documented in the literature, is also of interest:

$$H_{n+1}(x) = 2x H_n(x) - 2n H_{n-1}(x). \qquad (27.7)$$

As for any stationary state, the functions $\psi_n(ax)$ do not move, whatever the n value, showing nonzero values over the entire amplitude of the oscillator. Therefore, despite what is seen in Figure 27.2, the limit $n \to \infty$ on these states does not *actually* represent the classical limit. Yet states $\psi_z(x,t)$ can be defined, with arbitrary energy $\langle E \rangle \in [E_0, +\infty)$ $(E_0 = \hbar\omega/2)$, which *do move in a periodic way*, according to Schrödinger's equation, and reproduce *almost exactly* the motion of the classical oscillator.

The above can be achieved by taking a *superposition* state $\psi_z = \sum_n c_n \psi_n$, where the *moduli* $|c_n|^2$ of the coefficients follow a Poisson distribution $|c_n|^2 = P(\mu,n)$ with parameter $\mu = (\langle E \rangle - E_0)/\hbar\omega \equiv |z|^2$. Recall that the Poisson distribution is given by $P(\mu,n) = e^{-\mu}\mu^n/n!$. Hence the coefficients c_n are powers of a complex number $z = |z|e^{i\theta}$, $\theta \in [0, 2\pi]$, taking the form $c_n = e^{-|z|^2/2} z^n/\sqrt{n!}$. It is straightforward to see that z acquires the time dependence $z(t) = |z(0)|e^{i(\theta - \omega t)}$, just from Schrödinger's equation.

Such superposition states are called in the literature *coherent states* of the oscillator [1], and have many interesting properties that will not be detailed in this course. It can be seen, for instance, that $\Delta E = |z|\hbar\omega$. We shall simply note that, in the limit $|z|^2 \to \infty$, these states have an infinitely well defined energy, relative to $\langle E \rangle$, since $\Delta E/\langle E \rangle = \sqrt{\hbar\omega}\sqrt{\langle E \rangle - E_0}/\langle E \rangle \to 0$ as $\langle E \rangle \to \infty$. The ΔE result above follows from the well known expression of the standard deviation of the Poisson distribution $\sigma = \sqrt{\langle n \rangle} = \sqrt{\mu} = |z|$.

The harmonic oscillator potential can be extended to 3D, and the simplest version is considering that the characteristic frequencies are all equal in the 3 spatial directions: $\omega_x = \omega_y = \omega_z = \omega$. Such potential shows a central symmetry: $U(r) = (1/2)m\omega^2 r^2 = (1/2)m\omega^2(x^2 + y^2 + z^2)$, and the natural way to deal with it is solving the radial Schrödinger equation.

However, we see that the above potential has the unusual property of being separable in three identical terms for each cartesian coordinate. This offers the alternative way of a direct extension of the previous result in 1D, just by considering solutions in factorized form $\psi(r) = \psi_{n_1}(x)\psi_{n_2}(y)\psi_{n_3}(z)$.

Let us then write Schrödinger's equation:

$$\left[\frac{-\hbar^2}{2m}\left(\frac{\partial^2}{\partial x^2} + \frac{\partial^2}{\partial y^2} + \frac{\partial^2}{\partial z^2}\right) + \frac{1}{2}m\omega^2\left(x^2 + y^2 + z^2\right)\right]\psi(r) = E\psi(r),$$

which breaks up into three equations identical to the one studied above in 1D, with total energy defined by the sum: $E = E_{n_1} + E_{n_2} + E_{n_3}$. That is, the stationary states are in $1-1$ relationship with the sets of three integers (n_1, n_2, n_3). Note that all of them are allowed to be zero, taking the values $n_i = 0, 1, 2, \cdots \infty$.

Which allows us to immediately establish the *energy spectrum* of the 3D oscillator:

$$E = \left(n_1 + n_2 + n_3 + \frac{3}{2}\right)\hbar\omega = \left(n + \frac{3}{2}\right)\hbar\omega.$$

---

[1] see for example "Quantum Mechanics I" p. 149, A. Galindo and P. Pascual, Springer-Verlag, 1990.

In the above expression $n \equiv n_1 + n_2 + n_3$ is an integer number that, as for the 1D oscillator, takes the values: $n = 0, 1, \cdots \infty$. The wave functions are expressed in detail, using the Hermite polynomials, as:

$$\psi_{n_1 n_2 n_3}(r) = \prod_{i=1}^{3} N_{n_i} H_{n_i}(ax_i) e^{-a^2 x_i^2/2}.$$

We may use the Dirac notation in the form $|n_1 n_2 n_3\rangle$ to designate the above stationary states. Let us now see that, above the ground state $|000\rangle$ ($n = 0$), that is non degenerate, a *quantum degeneracy* g_n builds up that increases with increasing n. For example, for $n = 1$ the states: $|100\rangle$, $|010\rangle$ and $|001\rangle$ show equal energy. To calculate g_n, let us determine how many sets of 3 natural integers can be found with a fixed sum: $n = n_1 + n_2 + n_3$.

The problem is equivalent to another of combinatorial nature: in how many ways can we distribute n balls over 3 boxes, allowing for the boxes to be empty. Note that, once two boxes have been filled, the third one is determined by the difference. Therefore, $n + 2$ elements (2 possible holes) need to be distributed over 2 boxes, that is:

$$g_{\text{oscillator}} = \binom{n+2}{2} = \frac{1}{2}(n+1)(n+2). \qquad (27.8)$$

Similarly to the 1D case, each energy level has a well defined *parity*, as determined by the product: $(-1)^{n_1}(-1)^{n_2}(-1)^{n_3} = (-1)^n$, that alternates sign for increasing integers n.

We should bear in mind that, in the same way as the energy levels E_n are measurable, so are the degeneracy levels g_n, through the *entropy*, i.e. by measuring the specific heat of one mole of oscillators. Thus the previously obtained value of g_n, from our analysis in cartesian coordinates, *must* remain valid when the analysis is performed in spherical coordinates.

For its pedagogical interest, let us provide here some indications about the handling of the 3D harmonic oscillator in spherical coordinates [2]. When using the radial Schrödinger equation with potential $U(r) = (1/2)m\omega^2 r^2$, solutions must be found that are square-integrable in $\mathbb{R}^3$. Their behaviour as $u_l \propto r^{l+1}$ for $r \to 0$, shown in Chapter 17, and the fact that $u_l(r) \propto e^{-a^2 r^2/2}$ for $r \to \infty$, as observed in 1D, suggest the function change:

$$u_l(r) \equiv r^{l+1} e^{-a^2 r^2/2} v_l(r).$$

---

[2] a complete calculation, following a different path, but with the same notation as in the subsequent paragraphs, can be found in https://en.wikipedia.org/wiki/Particle_in_a_spherically_symmetric_potential.

Chapter 27. THE 3D HARMONIC OSCILLATOR

The resulting radial equation for $v_l(r)$ is simplified if we define the argument of the function to be $\rho = a^2 r^2$, with $w_l(\rho) = v_l(r)$, the parameter $a \equiv \sqrt{m\omega/\hbar}$ being the characteristic inverse length of the 1D oscillator. As seen in the cases of Hydrogen atom and 1D oscillator, a power series expansion is stated for the function w_l (in powers of ρ), that satisfies a recurrence relation law for the coefficients of the series. The requirement of being square-integrable is also met in this case by demanding that the series degenerates into a polynomial of degree n_r, with $n_r = 0, 1, 2, \cdots \infty$ defining the radial energy quantization.

Indeed the solutions for the stationary states, making reference to the above-mentioned polynomials, are expressed in compact form as:

$$\psi_{n_r, l, m_l} = N_{nl}\, r^l\, Y_l^{m_l}(\theta, \phi)\, e^{-r^2 a^2/2}\, L_{n_r}^{(l+\frac{1}{2})}(a^2 r^2),$$

where $L_{n_r}^{(l+\frac{1}{2})}(x)$ designate the *generalized Laguerre* polynomial of degree $n_r = (n-l)/2$ [3] evaluated on the argument $x = a^2 r^2$ (as will be seen, $n - l$ is an *even* integer).

The energy spectrum then becomes:

$$E_{n_r l} = \left(2 n_r + l + \frac{3}{2}\right)\hbar\omega, \qquad (27.9)$$

with $n_r = 0, 1, \cdots \infty$ (radial energy) and $l = 0, 1, \cdots \infty$ (angular momentum). As expected, the spectrum is identical to that found with cartesian coordinates, provided the identification $n = 2 n_r + l = n_1 + n_2 + n_3$ is made.

Let us focus on the Hilbert space of dimension g_n corresponding to a given energy level of the 3D harmonic oscillator. The above results provide an interesting example of two different orthonormal bases, that subtend the same space. Using Dirac notation, these bases are defined by the states: $\{|n_1 n_2 n_3\rangle\}$ and $\{|n_r l m_l\rangle\}$.

These two sets of physical states are completely different: in the first basis, the energy imprinted on each coordinate axis is well defined, but the ocillator's angular momentum is not. In the second basis, we have maximal information about the angular momentum, through the quantum numbers (l, m_l), but an energy measurement along a given coordinate direction (say the X-axis) will be subject to quantum fluctuations.

---

[3] polynomial that takes the form: $L_{n_r}^{(\alpha)}(x) = \sum_{k=0}^{n_r} (-1)^k \binom{n_r + \alpha}{n_r - k} x^k / k!$, where $\binom{m}{s} = \frac{m(m-1)\ldots(m-s+1)}{s!}$ are *generalized* binomial coeficients, with the constant: $N_{nl} = \left(\frac{2^{n+l+2} a^{2l+3}}{\sqrt{\pi}}\right)^{\frac{1}{2}} \left(\frac{\left(\frac{n-l}{2}\right)! \left(\frac{n+l}{2}\right)!}{(n+l+1)!}\right)^{\frac{1}{2}}$.

Every state of energy E of a 3D oscillator can be expressed as a linear combination of states of either of the two bases (a $g_n \times g_n$ rotation matrix exists that relates them). Only the presence of an additional term in the Hamiltonian of the oscillator can break the energy degeneracy, with the possible appearance of a new basis, with new quantum numbers (where the new Hamiltonian is *diagonal*). For example, the introduction of a strong magnetic field B with $H = -(\mu_B/\hbar)BL_z$ would cause the degeneracy to be partially broken, and states would then acquire slightly different energies according to the quantum number m_l, due to the Zeeman effect we saw in Chapter 18. The new basis is specifically $\{|n_r l m_l\rangle\}$, in this case.

Let us observe in detail the symmetry of the oscillator states of the basis $|n_r l m_l\rangle$. All states $|n_1 n_2 n_3\rangle$ have a well defined parity: $(-1)^n$, that is shared by the states $|n_r l m_l\rangle$, because these are linear combinations of the former. Given that each state $|n_r l m_l\rangle$ has parity $(-1)^l$, we must have $(-1)^n = (-1)^l$ at all energy levels. This is precisely the reason why the first term of equation $n = 2n_r + l$ was bound to be an *even* integer. Thus for a given n, the sequence of l values sharing the same energy, in the 3D oscillator, are: $l = n, n-2, n-4, \cdots 0$, or else: $l = n, n-2, n-4, \cdots 1$, depending on the parity of n. That is, only the orbitals $ns, nd, ng, \cdots$, or $np, nf, nh \cdots$, are allowed.

It is an easy exercise checking that, for individual values of n, the quantum degeneracy g_n calculated by formula (27.8), is confirmed in the basis $|n_r l m_l\rangle$, as it was to be expected. The energy degeneracy shown by the 3D oscillator, with respect to l values, is similar to the one we have seen for the Coulomb potential, and these two potentials are to be considered as mathematically singular, with regard to this property.

There is yet another important feature of the quantum oscillator, dealing with photon emission and absorption under Bohr's formula $E_n - E_m = \hbar\omega$. The point is that n and m cannot just be arbitrary integers, but they must meet the equation $m = n \pm 1$.

Let us realize, first of all, that a subtle coincidence happens for the harmonic oscillator, between two frequencies ω that should not a priori be coincidental, namely its *characteristic* frequency $\omega_{osc} = \sqrt{k/m}$, and the frequency of the emitted photon: $\hbar\omega_{photon}$. Bohr's formula implies that $\hbar\omega_{photon}$ must be an integer multiple of $\hbar\omega_{osc}$. However, the experimental data show that a photon is never emitted with frequency double, triple, $\cdots$ of its characteristic frequency.

To understand the above phenomenon, let us make a small incursion into quantum electrodynamics. In the quantum oscillation implicit in Bohr's formula (discussed in Chapter 13) the local charge excess at point r can be associated, at any instant of time when $\cos(\omega t - \phi_{nm}) = 1$ is fulfilled, with the vector $d = \langle \psi_n | 2er | \psi_m \rangle$. Its complex modulus $|d|$ represents the maximum distance the charge has been displaced along each coordinate axis $(|d_x|, |d_y|, |d_z|)$, once averaged over all $\mathbb{R}^3$ space. Indeed, the above integration over the azimuth ϕ only depends on the factors $e^{im_l \phi}$ of each spherical harmonic. Defining $M \equiv m_{l_m} - m_{l_n}$, it is easy to check that a nonzero integral is obtained only in *two* cases: $M = \pm 1$ and $M = 0$. In the first case, the electric dipole *rotates* dextro/levo in the horizontal plane (recall Figure (13.1)) with $d = d_x(1, \pm i, 0)$, and in the second case it *oscillates* over the Z-axis with $d = d_z(0, 0, 1)$ having $d_x, d_z \in \mathbb{R}$. Note the functions ψ_n and ψ_m must have opposite parity, to allow $d \neq 0$.

The *imaginary* factor $\pm i$ is a mathematical relic of the existence of a $\pm \pi/2$ phase shift between the Y-component and the X-component in the above-mentioned rotation, that is equivalent to replacing the function $\cos(\omega t \pm \phi)$ by the function $\sin(\omega t \pm \phi)$ (recall the time-dependent analysis performed when deriving formula (13.1)).

In such a virtual charged oscillation, of frequency ω, the acceleration a and the amplitude $A = |d_i|/e$ are actually related by expression $a = A\omega^2$, thus $e^2 a^2 = |d_i|^2 \omega^4$. Hence the conditions are met to apply Larmor's radiation law of electromagnetism (seen in Section 2.3) $\mathcal{P} = e^2 a^2/(6\pi\varepsilon_0 c^3)$, on each component, in order to determine the *photon emission rate* (inverse time s^{-1}) simply by dividing the radiated power (in W) by the energy of *one* photon $\hbar\omega$:

$$W_{nm} = \left(\frac{\omega^4}{3\pi\varepsilon_0 c^3 \hbar\omega} \right) |\langle \psi_n | er | \psi_m \rangle|^2 . \qquad (27.10)$$

In the above expression a factor $1/2$ has been applied, to account for the time-averaging of either the function $\cos^2(\omega t)$ or the function $\sin^2(\omega t)$. It should be emphasized that formula (27.10) is *not specific* to the harmonic oscillator case, but generic, in Quantum Electrodynamics, for the spontaneous radiation of every system having radial symmetry [4].

---

[4] a more profound treatment of spontaneous radiation, without recourse to Larmor's radiation formula (and having identical result) is achieved by representing the E- and B-fields with quantized operators, in the dipole approximation. The radiation then arises from the action of electromagnetic waves in a vaccuum, at Bohr's frequency. These waves impart the particle located at point r an energy eEr, where the orientation of the electric field E must be averaged. The photon phase-space then comes into play to obtain (27.10).

It is interesting to note that the product of coefficients of the *very weak* quantum superposition $\psi = c_n \psi_n + c_m \psi_m$ (with $|c_n|^2 + |c_m|^2 = 1$), that is inherent to the light emitted by spontaneous radiation seen in Chapter 13, *can be estimated* as: $|c_n c_m| = W_{nm}/\omega \ll 1$. This comes as a simple consequence of the uncertainty principle applied to the photon emission time Δt, as $\Delta E \Delta t \sim \hbar \iff |c_n c_m| \hbar \omega \Delta t \sim \hbar$, after taking $\Delta t = 1/W_{nm}$.

To understand now the aforementioned one-step photon jumps of the harmonic oscillator, let us first consider the 1D case, and take into account the recursive property of the Hermite polynomials expressed in the form $H_{n+1}(x) = 2xH_n(x) - 2nH_{n-1}(x)$, and the parity $(-1)^n$ of each polynomial $H_n(x)$. Indeed, we see that only the energy level $|n+1\rangle$ may decay to $|n\rangle$. The wave function of level $|n+2\rangle$ can always be descomposed as $H_{n+2}(x) = 2xH_{n+1}(x) - 2(n+1)H_n(x)$. The first term throws zero for x^2 being an even function $\langle n|2ex^2|n+1\rangle = 0$, $|n+1\rangle$ and $|n\rangle$ having opposite parity, and also the second term throws zero, because $\langle n|x|n\rangle = 0$. Likewise $\langle n|ex|n+3\rangle = 0$, $|n+3\rangle$ and $|n\rangle$ having equal parity, and so on for $|n+4\rangle$, $|n+5\rangle$, etc.

The result $m = n \pm 1$ is immediately confirmed in 3D, taking into account that $r = x\mathbf{i} + y\mathbf{j} + z\mathbf{k}$. Indeed, let us define $|n\rangle \equiv |n_1 n_2 n_3\rangle$. Thus the element $\langle n|ex|m\rangle$ equals the product $\langle n_1|ex|m_1\rangle \cdot \langle n_2|m_2\rangle \cdot \langle n_3|m_3\rangle$, which becomes non zero only if $m_1 = n_1 \pm 1$, according to the above result in 1D, and in addition $n_2 = m_2$ and $n_3 = m_3$, by the orthonormality of the H_{n_i}. By the same token, $\langle n|ey|m\rangle = 0$ and $\langle n|ez|m\rangle = 0$ are obtained. Obviously, the same reasoning applies for the y and z coordinates.

The broad implication of the harmonic oscillator in physics was earlier pointed out, in Section 2.4. As an example of 1D oscillators, we have all diatomic molecules, that are gases at room temperature. The molecule rotation is best understood as part of the 3D oscillator-rotor that we have just discussed. On the other hand, we find examples of intrinsically 3D oscillators in all solids and liquids that transmit sound with a given frequency ω. The above ideas constitute the groundwork to assess the internal energy and entropy of such systems, in statistical mechanics.

The heat capacity of a solid at very low temperatures was already established in 1912 by the Dutch physicist Peter Debye, who unveiled its quantum structure as a *phonon* gas residing in the solid, having bosonic character.

www.ingramcontent.com/pod-product-compliance
Lightning Source LLC
Chambersburg PA
CBHW082106220526
45472CB00009B/2061